ADAPTIVE SLIDING MODE NEURAL NETWORK CONTROL FOR NONLINEAR SYSTEMS

Emerging Methodologies and Applications in Modelling, Identification and Control

ADAPTIVE SLIDING MODE NEURAL NETWORK CONTROL FOR NONLINEAR SYSTEMS

Edited by

YANG LI

Lecturer, Institute of Information Science and Engineering, Hebei University of Science and Technology, Shijiazhuang, China

JIANHUA ZHANG

Lecturer, Institute of Electrical Engineering, Hebei University of Science and Technology, Shijiazhuang, China

QIONG WU

Lecturer of the School of English Language and Literature, Hebei University of Science and Technology, Shijiazhuang, China

Series Editor

QUAN MIN ZHU

ELSEVIER

ACADEMIC PRESS

An imprint of Elsevier

Academic Press is an imprint of Elsevier
125 London Wall, London EC2Y 5AS, United Kingdom
525 B Street, Suite 1650, San Diego, CA 92101, United States
50 Hampshire Street, 5th Floor, Cambridge, MA 02139, United States
The Boulevard, Langford Lane, Kidlington, Oxford OX5 1GB, United Kingdom

Library of Congress Cataloging-in-Publication Data

A catalog record for this book is available from the Library of Congress

British Library Cataloguing-in-Publication Data

A catalogue record for this book is available from the British Library

ISBN: 978-0-12-815372-7

For information on all Academic Press publications visit our website at https://www.elsevier.com/books-and-journals

Publisher: Mara Conner
Acquisition Editor: Sonnini R. Yura
Editorial Project Manager: John Leonard
Production Project Manager: R.Vijay Bharath
Cover Designer: Miles Hitchen

Typeset by TNQ Technologies

CONTENTS

AUTHOR BIOGRAPHIES

Yang Li

Yang Li is a lecturer at the Institute of Information Science and Engineering, Hebei University of Science and Technology, Shijiazhuang, China.

She obtained her Ph.D. from Yanshan University, China, in 2014. She was a visiting student of the University of the West of England in 2012. She is the reviewer of *International Journal of Modelling, Identification and Control*, *IET Control Theory & Applications*, and *Journal of Systems and Control Engineering*. She is the associate editor of "*International Journal of Applied Mathematics in Control Engineering.*" She is also a member of the editorial board of *American Journal of Applied Mathematics*. Her main research interests are in the areas of sliding mode control, neural network control, time delay system, and control applications. Email: liyang@hebust.edu.cn.

Jianhua Zhang

Jianhua Zhang is a lecturer at the Institute of Electrical Engineering, Hebei University of Science and Technology, Shijiazhuang, China.

He obtained his BS degree from Jilin Normal University, China, in 2003; MS degree from Yanshan University, China, in 2006; Ph.D. from Yanshan University, China, in 2011; and post doctoral degree in UWE, UK, in 2013. His main research interests are in the areas of nonlinear control systems, control systems design over network, and intelligent control. He is the acting Editor (and Founder) of *International Journal of Applied Mathematics in Control Engineering* and associate editor of *International Journal of Analytical Methodologies and Applications in Engineering*. He is also a member of the editorial board of *American Journal of Applied Mathematics*. Email: zhangjianhua@hebust.edu.cn.

Qiong Wu

Qiong Wu is a lecturer at the School of English Language and Literature, Hebei University of Science and Technology, Shijiazhuang, China.

She obtained her BA and MA degrees from Hebei University of Science and Technology, China, in 2010 and the University of Sheffield, England, in 2011, respectively. Her main research interests are in the areas of corpus linguistics, translation, and language teaching. Email: wuqiong@hebust.edu.cn.

PREFACE

It has been observed that classical approaches in control system design have been almost uniquely to put plant model and controller together to design controller parameters. Although this induces unnecessary complexity and repetition, it is still resolvable with linear plant models. However, it is very difficult to find solutions for nonlinear models, particularly in establishing general design procedures/framework, which is why commonly agreed nonlinear control system design has no distinct approach, it must be worked case by case. To reduce such difficulties, several methodologies have been proposed such as iterative learning control, piecewise linearization control, point-wise linearization, and U-model method. So, we need to build a new structure to describe the nonlinear systems stability analysis and control methods.

The following are major themes of the book: (1) the basic concepts of nonlinear systems stability analysis and classical nonlinear systems control method; (2) the stability analysis of complex nonlinear system; (3) the nonlinear systems controller design of complex nonlinear systems; and (4) some industrial application.

This book gives an introduction to basic nonlinear systems architectures and control methods. Emphasis is placed on the mathematical analysis of these systems, on methods of controlling them, and on their application to practical engineering problems in areas such as aircraft path planning.

The book is an outgrowth of our research experience in China and United Kingdom, so this text is primarily intended for applied mathematicians and control science and engineering researchers and engineers in China, India, and North America. A first-year graduate-level course on nonlinear systems or control in China, India, and North America may also use the book for self-study or reference. The authors provide practical examples throughout. However, the authors focus on the control theory and applications research in this book, could be suitable for any interested audience, with no geographic limitation.

The purpose of this book is to introduce the nonlinear systems basic knowledge, analysis and control methods, and applications in various fields. This book enables the audience to understand the basic architectures of control science and engineering as well as to aid their own specialized subjects with the help of the control theory they have learned. It also contains a wide range of nonlinear systems control methods which can help the audience master the system controllers design methods and open their minds to the various applications. The area of nonlinear system analysis and control methods has progressed rapidly in recent years and we aim at the most recent themes and topics of current research. This book can be a helpful resource to aid teaching the subject efficiently within limited course hours.

ACKNOWLEDGMENTS

This work was supported by Hebei Province Nature Fund Grant F2015208128.Project in Hebei Province Department of Education QN20140157, BJ2016020. I, hereby, gratefully acknowledge this support.

I wish to express my sincere appreciation to the following professors who have encouraged and guided me throughout this book: Prof. Quanmin Zhu, University of the West of England; Prof. Dongwen Zhang, Hebei University of Science and Technology; and Prof. Xueli Wu, Hebei University of Science and Technology.

I also wish to thank my parents, my husband, and my daughter for their loving considerations and great confidence in me all through these years.

Yang Li
2018/2/28

INTRODUCTION

AN OVERVIEW OF EACH CHAPTER

Man-made systems with advanced functionalities in practical industrial processes usually exhibit nonlinearities, time-varying unpredictable parameters, and other complexities from design limitation, operation conditions, integrated inherent dynamics, and even internal and external disturbances. In addition, delay is also a common phenomenon encountered in the actual industrial production process, which makes the control system design even challenging. Therefore the robust sliding mode control method for uncertain nonlinear systems with time delay has attracted greater attention. In the thesis, the problems of robust sliding mode control method for uncertain nonlinear systems with time delay are studied by using linear matrix inequality, adaptive backstepping and neural network, and robust control and finite-time control methods. A brief introduction of the major research contents and achievement is outlined in the following.

In Chapter 1, first, the stability analysis including Lyapunov stability and robust stability is studied. Second, it studies nonlinear design tools including adaptive control, sliding mode control, and U-model method.

In Chapter 2, first, nonlinear neural networks systems are studied, and the new exponential stability criterion is provided in the form of linear matrix inequality based on Lyapunov stability theory. Second, it studies nonlinear neutral systems with uncertain parameters and external disturbance, the new robust stability criterion is provided in the form of linear matrix inequality based on Lyapunov stability theory, and the system uncertainties are disposed by the matrix inequality technique.

In Chapter 3, the classical nonlinear systems control methods, such as feedback control and backstepping control are studied. First, for a class of nonlinear system, the feedback controller is designed based on Lyapunov stability theory in order to achieve asymptotic stability of the closed-loop systems. Second, backstepping controller is designed based on Lyapunov stability theory in order to make output of the system achieve the given reference signal.

In Chapter 4, first, the chaotic nonlinear systems are studied, and the super-twisting synchronization controller is designed based on the U-model method and Lyapunov stability theory in order to achieve synchronization between the master and slave systems. Second, it studies the super-twisting control algorithm that is used to analyze the nonlinear control system based on the U model. The nonaffine nonlinear systems are described, and the one-order and *n*-order nonlinear system is designed. Third, it studies discrete nonlinear systems with matching perturbations, the sliding function is proposed, the nonlinear sliding mode controller is designed based on Lyapunov stability theory, the controller parameters are obtained by matrix technique, and the controller input is obtained by the Newton–Raphson Algorithm in order to achieve uniformly asymptotic stability of the closed-loop system, meanwhile eliminating the chattering effect.

In Chapter 5, first, it studies a generalized procedure in identification and control of a class of time-varying-delayed nonlinear dynamic systems. Under the framework, recurrent neural network is developed to accommodate the online identification, which the weights of the neural network have iteratively and adaptively updated through the model errors. Then indirect adaptive controller is designed based on the adaptive law of the controller, which is designed for the controller design. Second, it studies a neural network–based finite-time adaptive controller for a class of uncertain nonaffine nonlinear systems in pure feedback prototype. In this framework, homeomorphism mapping and finite-time convergence are introduced to ensure the state constraint and convergence in finite time, respectively. To prevent constraint violation, the homeomorphism mapping is applied to a chosen compact superset, which contains initial condition and virtual superset without constraint, and then the finite-time converge method is used to ensure converge time after the mapping transform. Accordingly, it provides a generic prototype for finite-time adaptive law to train neural networks.

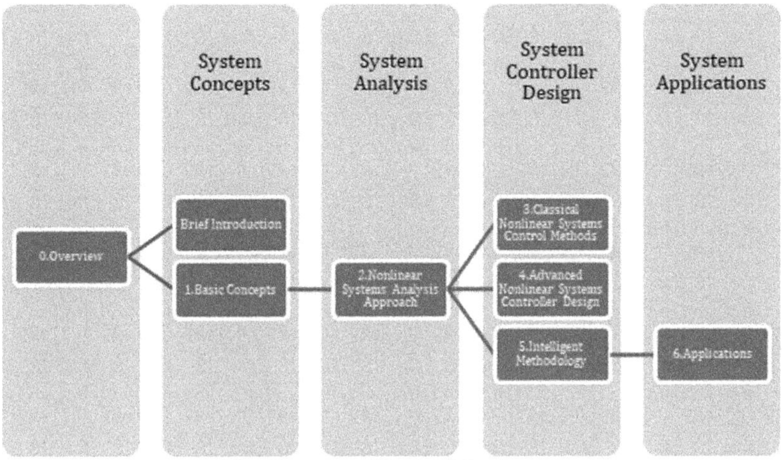

The Structure Chart of the book.

In Chapter 6, first, it studies the intelligent flight task algorithm for aircraft to effectively determine and search the best flight routes and complicated conditions, which include complicated topography and threats from other flights in the route airspace. Second, it studies the multilateration system theory and positioning algorithm using backpropagation network.

CHAPTER 1

Basic Concepts

Contents

Adaptive Sliding Mode Neural Network Control for Nonlinear Systems
ISBN 978-0-12-815372-7
https://doi.org/10.1016/B978-0-12-815372-7.00001-X

First, Chapter 1 studies stability analysis, including Lyapunov stability and robust stability. Second, it studies nonlinear design tools, including adaptive control, sliding mode control, and U-model method (Fig. 1.1).

1.1 LYAPUNOV STABILITY

1.1.1 Lyapunov Stability Theory

1.1.1.1 Introduction

1. Overview

   Lyapunov stability theory was come out of Lyapunov, a Russian mathematician in 1892, and came from his doctoral dissertation. Until now, the theory of Lyapunov stability is still the main theoretical basis of almost all system-controller design (Chen, 1984).

2. Function

   First of all, the Lyapunov stability theory is understood through the picture. The control of the trajectory of the space transports is based on the Lyapunov stability theory.

   The Lyapunov stability theory is used to describe the stability of a dynamic system (Fig. 1.2).

3. Application

   The application of this theory to control is mainly based on the Lyapunov stability theorem. There are two methods

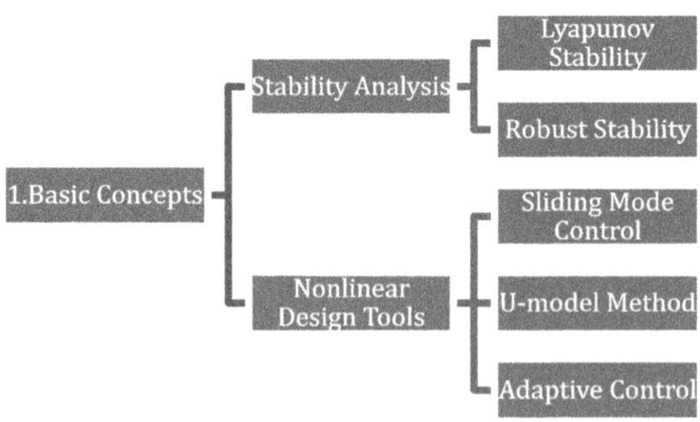

Figure 1.1 The structure chart of Chapter 1.

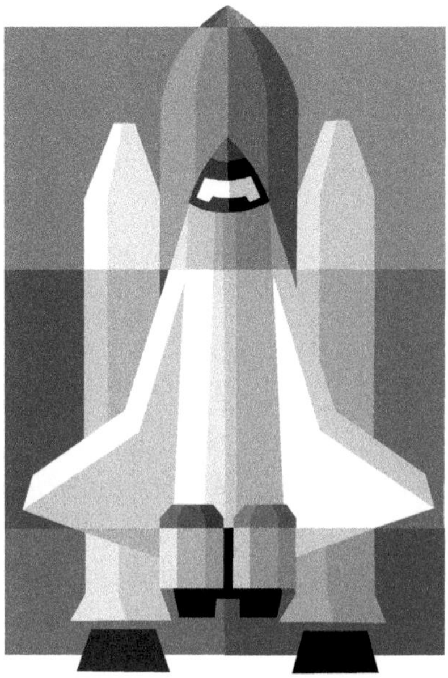

Figure 1.2 Trajectory of the missile.

for specific application, and the popular one is the Lyapunov second method.

1.1.1.2 Autonomous System

1. Definition

   The autonomous system means no external input system (Khalil, 2002). The autonomous system is described as a form of differential equations $\dot{x} = f(x, t),\ \ x(t_0) = x_0 t \in [t_0, \infty)$, where x is the system state, f is the nonlinear function with x and t, t is time.

2. Example

   For example, there is a bicycle there; suddenly, the bicycle fell down by a gust of wind, the wind is the external input, but the bike has also been affected by gravity; it is not entirely without external input, so there is no external input is a relative concept (Fig. 1.3).

   (Therefore, the autonomous system is a relative concept.)

Figure 1.3 Example of the autonomous system.

1.1.1.3 Equilibrium State

1. Definition

   The state of equilibrium is the state of keeping the system still in the absence of external interference. If there is a state x_e, for any t great than t_0, a set of derivatives of x_e satisfied zero, then it is called the equilibrium state or equilibrium point $\dot{x}_e = f(x_e, t) = 0, \forall t \in [t_0, \infty)$ (Dazhong, 2002).

2. Example (Fig. 1.4)

   There are two balls, assuming that two balls are in this position and remain motionless. In the absence of external interference, we do not move it; the two balls will continue to remain stationary. If the displacement of the ball is regarded as a state x_e, the derivative of the apparent state satisfies this condition $\dot{x}_e = f(x_e, t) = 0$, so the two balls are all in equilibrium.

   (As we can see from this example, the equilibrium state is not only one. For example, the position of the two balls is in the equilibrium position, and the two points are all the equilibrium points of the system.)

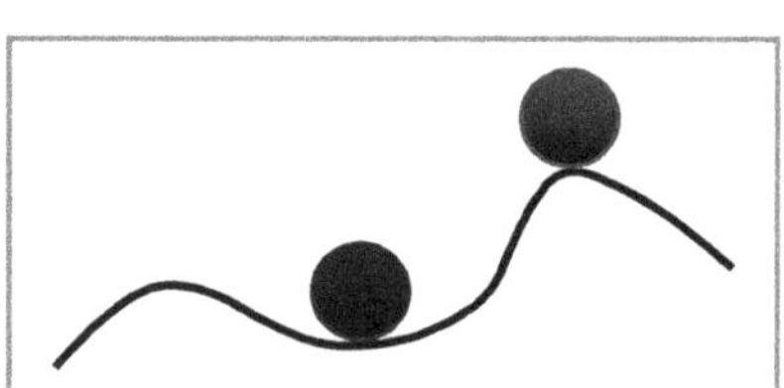

Figure 1.4 Example of equilibrium state.

1.1.1.4 Lyapunov Stability

1. Concept of Limit in Higher Mathematics

   When the independent variable tends to infinity, the function $f(x)$ value is located in a small area near the limit value, $\forall \varepsilon > 0 \exists X > 0$ when $|x| > X$, we have $|f(x) - A| < \varepsilon$; the concept of stability in the sense of Lyapunov is similar to the concept of limit in higher mathematics.

2. Definition of Lyapunov Stability

   a. The geometric interpretation of Lyapunov stability (Fig. 1.5)

      The initial state x_0 is assumed to be located in a closed ball region with equilibrium state x_e of its center and the arbitrary δ of its radius; when time t tends to infinity, the system state can be located in the closed ball with equilibrium state x_e as center, any with arbitrary ε as radius, then the system is in Lyapunov stability.

- The system trajectory: the center x_e, and closed ball with radius δ and ε, respectively, called $s(\delta)$ and $s(\varepsilon)$. At any point in $s(\delta)$, the trajectory of the system will not run out of $s(\varepsilon)$, this is defined by the definition. But the system trajectory does not necessarily converge to the equilibrium state x_e, it may not even fall into $s(\delta)$, that may stay any point outside

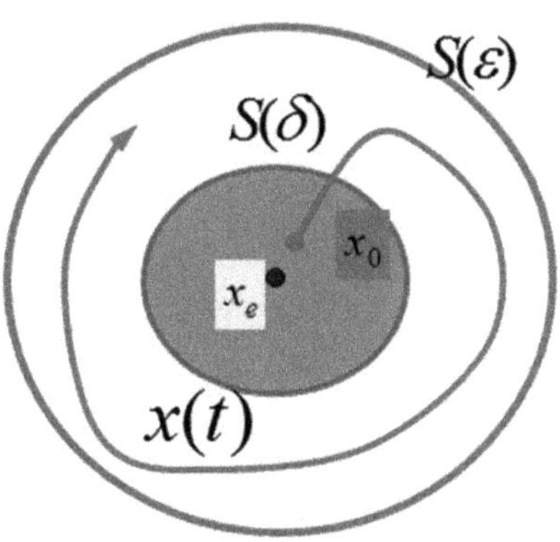

Figure 1.5 Definition of Lyapunov stability.

$s(\delta)$, but it may also be spinning outside $s(\delta)$, but they are all stable.

- The system parameters: when determining the stability of the system, the definition of ε is first determined, and after it is determined, it will find one δ corresponding to it.
- The Mathematical Description of Lyapunov Stability

  The equilibrium state $x_e = 0$ of the autonomous system $\dot{x} = f(x, t)$ is said Lyapunov stability at t_0, if for any $\varepsilon > 0$, there exist $\delta(\varepsilon, t_0) > 0$ depending on ε and t_0 that satisfies the following inequality:

$$\|x_0 - x_e\| \leq \delta(\varepsilon, t_0)$$

  The disturbed motion $\phi(t; x_0, t_0)$ of any initial state x_0 satisfies the inequality:

$$\|\phi(t; x_0, t_0) - x_e\| \leq \varepsilon, \forall t \geq t_0$$

3. Example of Lyapunov Stability (Fig. 1.6)

Once the pendulum is subjected to external force, it will deviate from the equilibrium position. The displacement of a pendulum is regarded as a system trajectory, and the moment of external force loss is taken as the initial value t_0. When the initial state x_0 always falls into the closed area of the radius δ, the system trajectory still falls into the closed area with the radius ε, which satisfies the Lyapunov stability definition. So the single pendulum is stable in the sense of Lyapunov.

(The object of motion may also be Lyapunov Stable.)

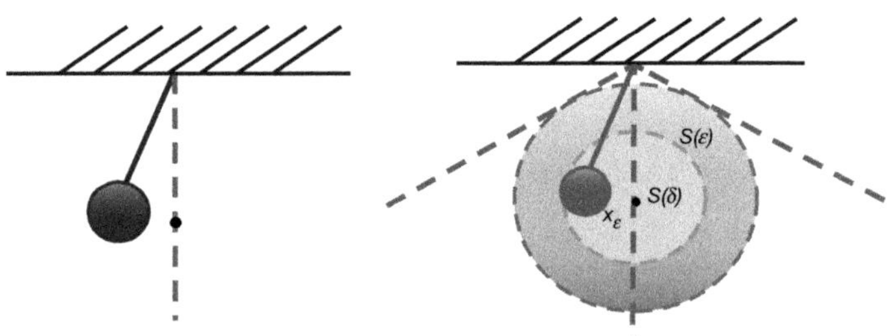

Figure 1.6 Example of Lyapunov stability.

1.1.2 Lyapunov Asymptotic Stability

1.1.2.1 Definition of Lyapunov Asymptotic Stability

1. The geometric interpretation of Lyapunov Asymptotic Stability (Fig. 1.7)

   The initial state x_0 is assumed to be located in a closed ball region with equilibrium state x_e of its center and the arbitrary δ of its radius, when time t tends to infinity, the system state can be converged to equilibrium point x_e, then the system is in Lyapunov asymptotic stability.

2. The Mathematical Description of Lyapunov Asymptotic Stability

   The equilibrium state $x_e = 0$ of the autonomous system $\dot{x} = f(x, t)$ is said to be at Lyapunov asymptotic stability at t_0, if

   - the disturbed motion $\phi(t;x_0,t_0)$, starting from any initial state $x_0 \in S(\delta)$, is bounded by the equilibrium state $x_e = 0$ for all $t \in [t_0, \infty)$;
   - the disturbed motion is asymptotically relative to the equilibrium state 1, that is,

   $$\lim_{t \to \infty} \phi(t; x_0, t_0) = x_e, \ \forall x_0 \in S(\delta)$$

3. The relationship between Lyapunov Stability and Lyapunov Asymptotic Stability

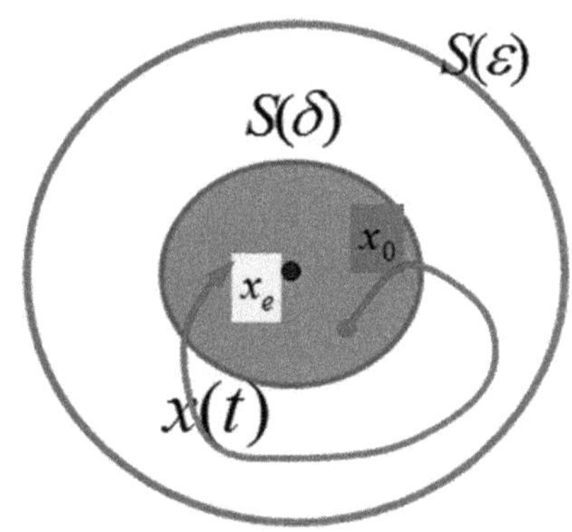

Figure 1.7 Definition of Lyapunov asymptotic stability.

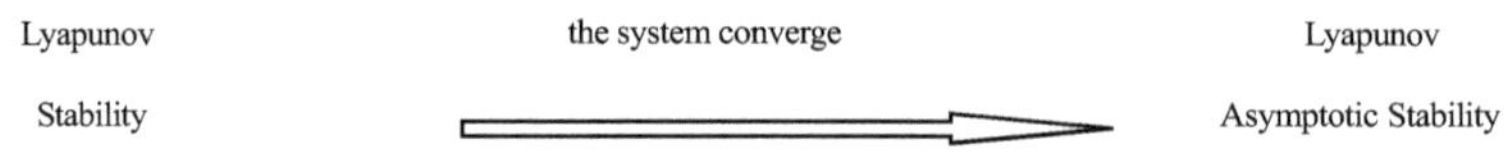

1.1.2.2 Example of Lyapunov Asymptotic Stability

Once the pendulum is subjected to external force, it will deviate from the equilibrium position. The displacement of a pendulum is regarded as a system trajectory, and the moment of external force loss is taken as the initial value t_0. When the initial state x_0 always falls into the closed area of the radius δ, the system trajectory still falls into the closed area with the radius ε, which satisfies the Lyapunov stability definition. So the single pendulum is stable in the sense of Lyapunov. But when a pendulum with energy loss is under the action of resistance such as friction, the trajectory of system $x(t)$ will eventually converge to the equilibrium state of x_e, which satisfies the asymptotic stability in the sense of Lyapunov (Fig. 1.8).

(It can be seen from this example the system is Lyapunov asymptotically stability first of all the system is Lyapunov stability.)

1.1.3 Lyapunov Uniform Asymptotic Stability

1.1.3.1 Definition of Lyapunov Uniform Asymptotic Stability

In the definition of Lyapunov asymptotic stability in the sense if the system trajectory at any time does not run out of $S(\varepsilon)$ and

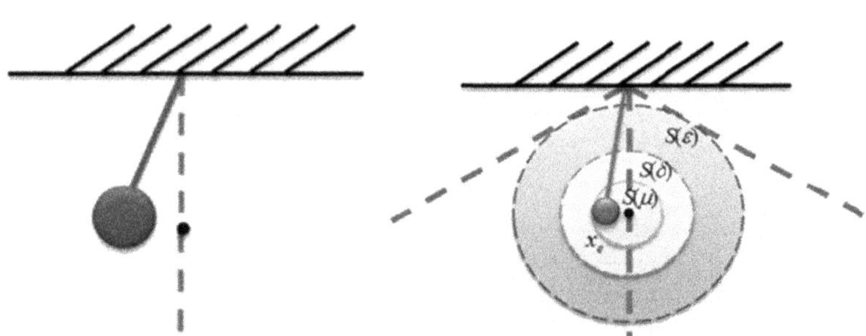

Figure 1.8 Example of Lyapunov asymptotic stability.

eventually converges to the equilibrium state x_e, the system is in Lyapunov uniform asymptotic stability.

1.1.3.2 The Relationship Between Lyapunov Asymptotic Stability and Lyapunov Uniform Asymptotic Stability

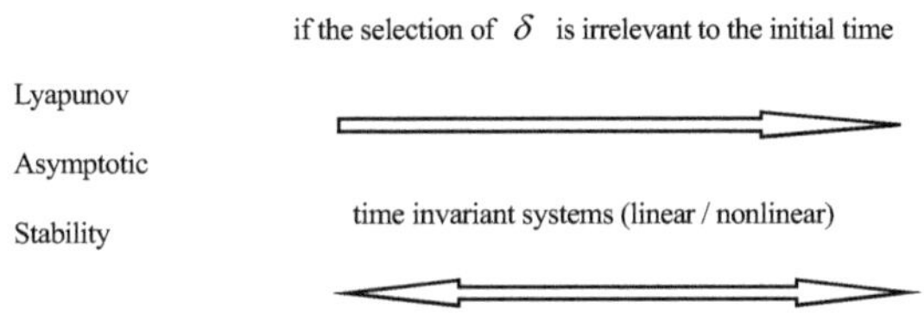

1.1.4 Lyapunov Global Asymptotic Stability

1.1.4.1 Definition of Lyapunov Global Asymptotic Stability

In the definition of Lyapunov asymptotic stability, if the system trajectories starting from any location do not run out of $S(\varepsilon)$ and eventually converge to the equilibrium state x_e, the system is in Lyapunov global asymptotic stability.

1.1.4.2 The Relationship Between Lyapunov Asymptotic Stability and Lyapunov Global Asymptotic Stability

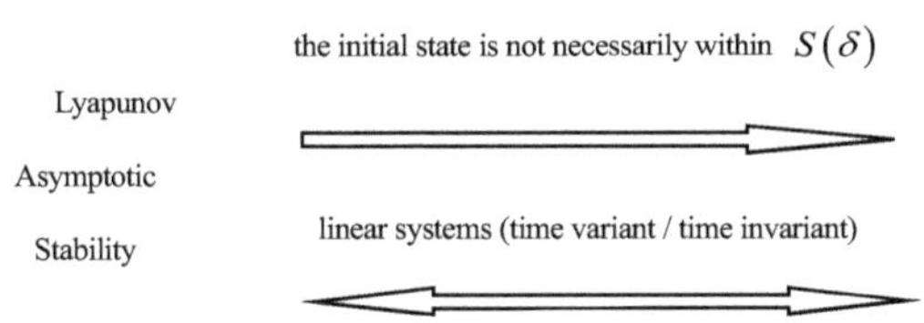

1.1.5 Lyapunov Instability

1.1.5.1 The Geometric Interpretation of Definition of Lyapunov Instability

The initial state is assumed that the x_0 is located in the center of the equilibrium state x_e, arbitrary radius δ closed ball, when time t tends to infinity, no matter how big $S(\varepsilon)$ or how small $S(\delta)$, the

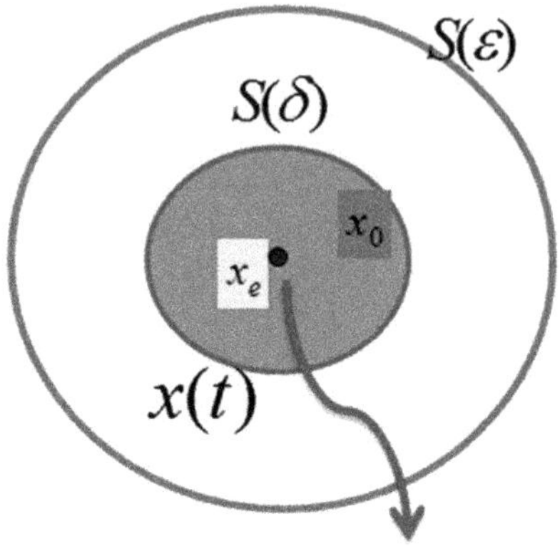

Figure 1.9 Definition of Lyapunov instability.

system state cannot be located in the center of the equilibrium state x_e, arbitrary radius ε closed ball, then the system is in Lyapunov instability (Fig. 1.9).

1.1.5.2 The Mathematical Description of Lyapunov Instability

The equilibrium state $x_e = 0$ of the autonomous system $\dot{x} = f(x, t)$ is said Lyapunov instability at t_0, if for any $\varepsilon > 0$, there nonexist $\delta(\varepsilon, t_0) > 0$ depending on ε and t_0 that satisfies the following inequality:

$$\|x_0 - x_e\| \leq \delta(\varepsilon, t_0)$$

The disturbed motion $\phi(t; x_0, t_0)$ of any initial state x_0 satisfies the inequality:

$$\|\phi(t; x_0, t_0) - x_e\| \leq \varepsilon, \forall t \geq t_0$$

1.1.6 Positive Definite Function

1.1.6.1 Definition

$V(x)$ is a positive definite function, if the following conditions are satisfied (Guangren, 2004).

1. $V(x)$ about x continuous derivable;
2. the function value of the origin is zero, that is $V(0) = 0$ $V(0) = 0$;

3. the function value of any point is positive definite, that is, to arbitrarily $x \neq 0$, we have $V(x) > 0$.
   Assume $\lim_{\|x\| \to \infty} V(\|x\|) = \infty$, then the positive definite

function $V(x)$ has the property of infinity.

1.1.6.2 The Relationship Between Positive Definite Function and Negative Definite Function

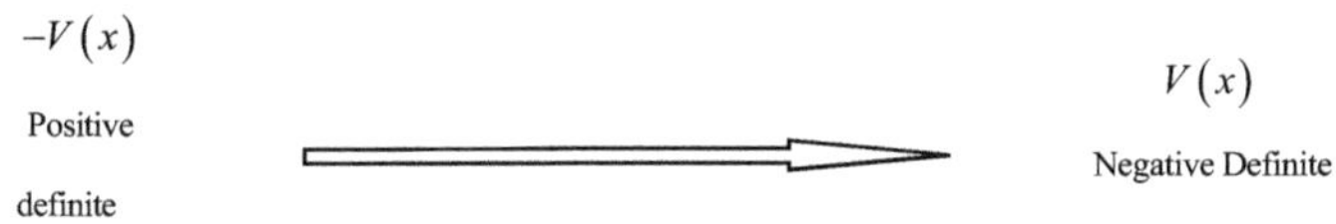

1.1.6.3 Example of Positive Definite Function

1. $V(x) = x_1^2 + x_2^2$ For all the components of a vector x, there is a continuous partial derivative of $V(x)$; when $x = 0$, $V(0) = 0$ for any $x \neq 0$, we have $V(x) > 0$, so $V(x)$ is positive definite function.
2. $V(x) = (x_1 + x_2)^2$ For all the components of a vector x, there is a continuous partial derivative of $V(x)$; when $x = 0, V(0) = 0$ but not for any $x \neq 0$, we have $V(x) > 0$, for example, when $x_1 = -x_2$, we have $V(x) = 0$, so $V(x)$ is not positive definite function and is semipositive definite function.
3. $V(x) = -x_1^2 - x_2^2$ $V(x)$ inverse get $-V(x) = x_1^2 + x_2^2$, for all the components of a vector x, there is a continuous partial derivative of $-V(x)$; when $x = 0$, we have $V(0) = 0$, for any $x \neq 0$, we have $V(x) > 0$, so $-V(x)$ is positive definite function, and then $V(x)$ is negative definite function.
4. $V(x) = x_1 x_2 + x_2^2$ For all the components of a vector x, there is a continuous partial derivative of $V(x)$; when $x = 0$, $V(0) = 0$, but for any $x \neq 0$, the symbols of $V(x)$ cannot be determined, so $V(x)$ is not positive definite function or negative definite function, it can be called indeterminate function.

(The function of the addition of the quadratic form conforms to the definition of the positive definite function.)

1.1.7 Lyapunov Function

1.1.7.1 Definition of Lyapunov Function

The core idea of Lyapunov second method is to set up a function similar to "energy" function, transform the problem of whether the system is stable into analyze the positive definite problem of the "energy" function and its first derivative, such "energy" function is Lyapunov function.

1.1.7.2 Construction of Lyapunov function

The construction of Lyapunov function is a complex problem. For a simple system, a class of Lyapunov function plays an important role in the stability analysis, that is, the quadratic form function.

$$V(x) = x^T Px = \begin{bmatrix} x_1 & x_2 & \cdots & x_n \end{bmatrix} \begin{bmatrix} p_{11} & p_{12} & \cdots & p_{1n} \\ p_{12} & p_{22} & \cdots & p_{2n} \\ \vdots & \vdots & & \vdots \\ p_{1n} & p_{2n} & \cdots & p_{nn} \end{bmatrix} \begin{bmatrix} x_1 \\ x_2 \\ \vdots \\ x_n \end{bmatrix}$$

Note that x is real vector and P is real symmetric matrix.

$$\begin{aligned} V(x) = x^T Px &= p_{11}x_1^2 + 2p_{12}x_1x_2 + \cdots 2p_{1n}x_1x_n \\ &+ p_{22}x_2^2 + 2p_{23}x_2x_3 + \cdots 2p_{2n}x_2x_n + \cdots + p_{nn}x_n^2 \end{aligned}$$

1.1.8 Lyapunov Stability Theorem and Lyapunov Global Uniform Asymptotic Stability Theorem

For continuous time, nonlinear, time-varying, autonomous system $\dot{x} = f(x, t)$, $t \in [t_0, \infty)$, if we can construct a scalar function $V(x,t)$, $V(0,t) = 0$ that has a continuous first-order derivative to x and t, and all nonzero state x in the state space $\Re^n$ are satisfied:

1. $V(x,t)$ is positive definite function, which means there are two continuous nonscalar functions $\alpha(\|x\|)$ and $\beta(\|x\|)$, where $\alpha(0) = 0$ and $\beta(0) = 0$, for all $t \in [t_0, \infty)$ and $x \neq 0$ satisfies
$$0 < \alpha(\|x\|) \leq V(x,t) \leq \beta(\|x\|)$$
2. $V(x,t)$ to t of the time derivative $\dot{V}(x,t)$ negative definite, that is there is a continuous nondecreasing scaler function $\gamma(\|x\|)$, where $\gamma(0) = 0$, for all $t \in [t_0, \infty)$ and $x \neq 0$ satisfies
$$\dot{V}(x,t) \leq \gamma(\|x\|) < 0$$
3. When $\|x\| \to \infty$, we have $\alpha(\|x\|) \to \infty$, that is $V(x,t) \to \infty$.

The equilibrium state $x = 0$ of the system is globally uniformly asymptotically stable.

1.1.8.1 Lyapunov Local Uniform Asymptotic Stability Theorem

For continuous time, nonlinear, time-varying, autonomous system $\dot{x} = f(x,t)$, $t \in [t_0, \infty)$, if we can construct a scalar function $V(x,t)$, $V(0,t) = 0$ that has a continuous first-order derivative to x and t and an attraction area around the origin of the state space Ω, make all nonzero state $x \in \Omega$ and $t \in [t_0, \infty]$ satisfies

1. $V(x,t)$ is positive definite function; and
2. $\dot{V}(x,t) \triangleq dV(x,t)/dt$ is negative definite function.

The equilibrium state $x = 0$ of the system is local uniformly asymptotically stable of the space Ω.

1.1.8.2 Example of Lyapunov Stability Theorem

Example: Consider the following system, choose the following Lyapunov function:

1. Determine the stability of the system according to the theorem

2. Draw the state trajectory of the system according to the solution of the system.

System Description:

$$\dot{x}_1 = 0.5x_1$$
$$\dot{x}_2 = 0.5x_2$$

Lyapunov Function and its derivative:

$$V(t,x) = e^{-2t}\left(x_1^2 + x_2^2\right)$$
$$V(t,0) = 0,\ V(t,x) > 0$$
$$\dot{V}(t,x) = -e^{-2t}\left(x_1^2 + x_2^2\right) \le 0$$

System Solution:

$$x_1(t) = x_1(t_0)\exp\left(\frac{1}{2}t - \frac{1}{2}t_0\right)$$
$$x_2(t) = x_2(t_0)\exp\left(\frac{1}{2}t - \frac{1}{2}t_0\right)$$

Remark: The given Lyapunov function and its derivative seem to conform to the conditions in the theorem, but the system is not stable from the system trajectory. What is the reason for this? In fact, careful thinking is not difficult to conclude that the given $V(t,x) = e^{-2t}\left(x_1^2 + x_2^2\right)$ is larger than zero, but the time-varying function also requires a bounded Lyapunov, which is greater than zero. Therefore, it is not a Lyapunov function. (Fig. 1.10) (The positive definite and bounded of the Lyapunov function is not a simple number greater than zero in a number.)

1.1.9 Robust Stability

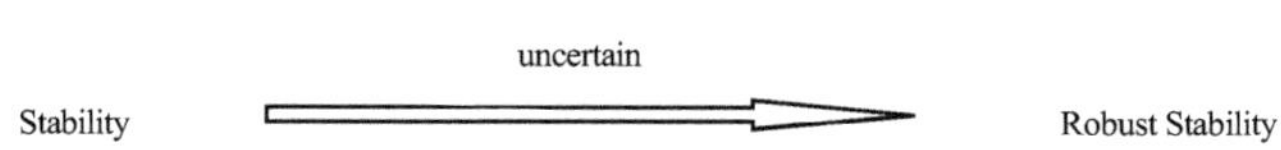

1.2 ADVANCED NONLINEAR SYSTEMS CONTROL

The controlled system can be divided into linear systems and nonlinear systems. Lyapunov stability analysis theory is suitable for both linear systems and nonlinear systems (Chen, 1984). Nonlinear system control method including sliding mode control, backstepping control, adaptive control, neural network control, and so on. In this book, it studies nonlinear system

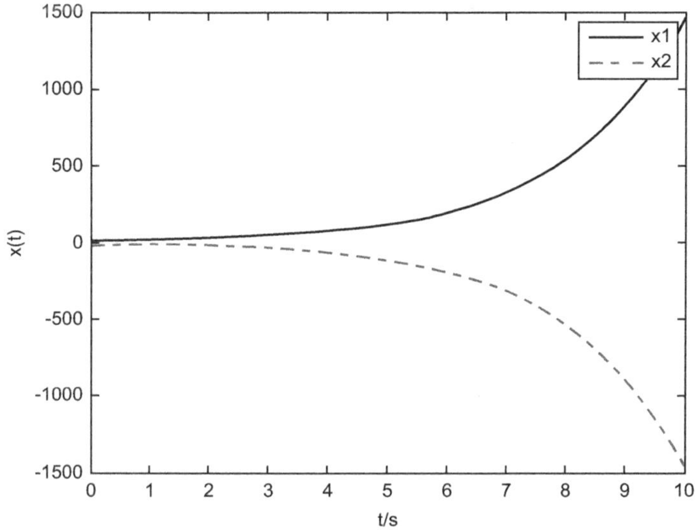

Figure 1.10 The state trajectory of the system.

control with uncertain and time delay using sliding mode control, backstepping control, adaptive control, and neural network control methods (Meng, 2018).

1.3 INTELLIGENT METHODOLOGY

Intelligent control is a class of control techniques that use various artificial intelligence computing approaches like neural networks, Bayesian probability, fuzzy logic, machine learning, evolutionary computation, and genetic algorithms (Åstr and Mcavoy, 1992; Wenxia Du et al., 2018; Shengkai Liu et al., 2018). In this book, it studies the nonlinear, time delay system identification and control using neural network, adaptive technique.

REFERENCES

Åstr, M.K.J., Mcavoy, T.J., 1992. Intelligent control. Journal of Process Control 2, 115–127.

Chen, C.T., 1984. Linear System Theory and Design. Holt, Rinehart and Winston.

Dazhong, Z., 2002. Linear System Theory. Tsinghua University Press, Beijing.

Guangren, D., 2004. Linear System Theory. Harbin Institute of Technology Press, Harbin.

Khalil, H., 2002. Nonlinear Systems, third ed.

Meng, R., 2018. Adaptive parameter estimation for multivariable nonlinear CARMA systems. International Journal of Applied Mathematical and Computer Modelling 1, 96–102.

Shengkai Liu, T.S., Wang, B., Jin, X., PENG, S., 2018. Multi-imu data fusion for indoor navigation. International Journal of Applied Mathematical and Computer Modelling 1, 15–22.

Wenxia Du, X.Z., Hailian, Du, Feng, L.V., 2018. Fault diagnosis of analog circuit based on extreme learning machine. International Journal of Applied Mathematical and Computer Modelling 1, 9–14.

CHAPTER 2

Nonlinear Systems Analysis Approach

Contents

This chapter first studies the nonlinear neural networks systems, and the new exponential stability criterion is provided in the form of linear matrix inequality (LMI) based on the Lyapunov stability theory. Second, it studies the nonlinear neutral systems with uncertain parameters and external disturbance, and the new robust stability criterion is provided in the form of LMI based on the Lyapunov stability theory and the system uncertainties are disposed by matrix inequality technique. Third, it studies neural network identification, a generalized procedure in the identification and control of a class of time-varying-delayed nonlinear dynamic systems. Under the framework, recurrent neural network is developed to accommodate the online identification, where the weights of the neural network are iteratively and adaptively updated through the model errors (Fig. 2.1).

Adaptive Sliding Mode Neural Network Control for Nonlinear Systems
ISBN 978-0-12-815372-7
https://doi.org/10.1016/B978-0-12-815372-7.00002-1

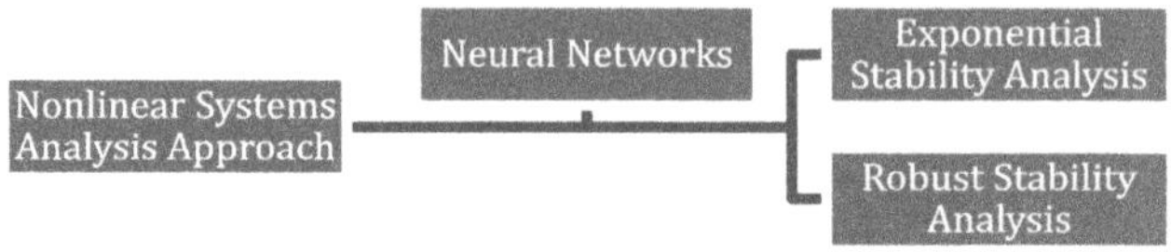

Figure 2.1 The structure chart of this chapter.

2.1 EXPONENTIAL STABILITY ANALYSIS OF CELLULAR NEURAL NETWORKS BASED ON LINEAR MATRIX INEQUALITY

2.1.1 System Formulation and Preliminaries

Consider the following neural network system with time delay:

$$\dot{y}(t) = -Cy(t) + Ag(y(t)) + Bg(y(t-\tau(t))) + S \tag{2.1}$$

where $y(\cdot) = [y_1(\cdot), y_2(\cdot), \ldots, y_n(\cdot)]^T$ is the state vector; $g(y(\cdot)) = [g_1(y_1(\cdot)), g_2(y_2(\cdot)), \ldots, g_n(y_n(\cdot))]^T$ denotes the neuron activation function; $S = [s_1, s_2, \ldots, s_n]^T$ is a constant input vector; C is a diagonal matrix with $C = diag[\,c_1 \quad c_2 \quad \ldots \quad c_n\,]$, $c_i > 0$, $1 \le i \le n$; A and B are the connection weight matrix and the delayed connection weight matrix, respectively; and $\tau(t)$ is the time-varying delay (Haoran He, 2018).

Assumption 1: Assume that the neuron activation function and the time-varying delay satisfies the following condition (Wu, 2018):

1. $g(0) = 0, \quad |g(\xi)| \le M$
2. $0 \le g(x) - g(y) \le l_i(x - y), \forall x, y \in R, L = diag(l_1, l_2, \ldots, l_n)$
3. $0 \le \tau(t) \le \tau, \quad \dot{\tau}(t) \le \mu < 1$

Assuming that there is an equilibrium point x^* for the neural network system (2.1), we have

$$\dot{y}^*(t) = -Cy^*(t) + Af\left(y^*(t)\right) + Bf\left(y^*(t-\tau(t))\right) + S \tag{2.2}$$

By the transformation $x(\cdot) = y(\cdot) - y^*$, the equilibrium point can be shifted to the origin, and consequently, system (2.1) can be transferred as

$$\dot{x}(t) = -Cx(t) + Af(x(t)) + Bf(x(t-\tau(t))) \tag{2.3}$$

where $x(\cdot) = [x_1(\cdot),\ x_2(\cdot), \ldots x_n(\cdot)]^T$ $f(x(\cdot)) = [f_1(x(\cdot)), f_2(x(\cdot)), \ldots f_n(x(\cdot))]^T$ $f_i(x_i(\cdot)) = g_i(x_i(\cdot) + y_i^*) - g_i(y_i^*)$.

The function $f_i(x_i(\cdot))$ satisfies that

$$f_i^2(x_i(\cdot)) \le L_i x_i(\cdot) f_i(x_i(\cdot)) \tag{2.4}$$

where $f_i(0) = 0, \quad i = 1, 2, \ldots n$.

2.1.2 Main Results

Theorem 2.1: System (2.3) is global exponential stable if there exist the symmetrical positive matrices P, Q and the positive matrices D, D_1, D_2, D_3, D_4, D_5, D_6, H, T_1, T_2, and L, where L is determined by the active function, such that the following LMI $\Xi < 0$ holds:

$$\Xi = \begin{bmatrix} \Xi_{11} & 0 & 0 & \Xi_{14} & 0 & \Xi_{16} \\ * & \Xi_{22} & 0 & 0 & \Xi_{25} & 0 \\ * & * & \Xi_{33} & 0 & 0 & \Xi_{36} \\ * & * & * & \Xi_{44} & 0 & \Xi_{46} \\ * & * & * & * & \Xi_{55} & 0 \\ * & * & * & * & * & \Xi_{66} \end{bmatrix} \tag{2.5}$$

where

$$\Xi_{11} = -PC - CP + 2\beta P - LHC - CHL + 2\beta HL + 2\beta LH + Q_{11} + R_{11} + LT_1L + LT_2L,$$

$$\Xi_{14} = PA - CD + Q_{12} + R_{12} + LHA - CH - 2\beta D - 2\beta H,$$

$$\Xi_{16} = PB + LHB,$$

$$\Xi_{22} = -e^{-2\beta\tau} R_{11} - e^{-2\beta\tau} LT_2L,$$

$$\Xi_{25} = -e^{-2\beta\tau} R_{12},$$

$$\Xi_{33} = -(1-\mu)e^{-2\beta\tau} Q_{11} - (1-\mu)e^{-2\beta\tau} LT_1L,$$

$$\Xi_{36} = -(1-\mu)e^{-2\beta\tau} Q_{12},$$

$$\Xi_{44} = DA + A^TD + Q_{22} + R_{22} - HA - A^TH - T_1 - T_2,$$

$$\Xi_{46} = DB - HB,$$

$$\Xi_{55} = -e^{-2\beta\tau} R_{22} + e^{-2\beta\tau} T_2,$$

$$\Xi_{66} = -(1-\mu)e^{-2\beta\tau} Q_{22} + (1-\mu)e^{-2\beta\tau} T_1.$$

Proof: Choose the following Lyapunov–Krasovskii functional (LKF) candidate:

$$V(t) = V_1(t) + V_2(t) + V_3(t) + V_4(t) \tag{2.6}$$

$$V_1(t) = x^T(t)Px(t) \tag{2.7}$$

where $P = P^T > 0$.

The derivative of $V_1(t)$ is

$$\begin{aligned}\dot{V}_1(t) &= 2x^T(t)P\dot{x}(t)\\ &= -2x^T(t)PCx(t) + 2x^T(t)PAf(x(t)) + 2x^T(t)PBf(x(t-\tau(t)))\end{aligned} \tag{2.8}$$

Then we have

$$\begin{aligned}\dot{V}_1(t) + 2\beta V_1 = &-2x^T(t)PCx(t) + 2x^T(t)PAf(x(t))\\ &+ 2x^T(t)PBf(x(t-\tau(t))) + 2\beta x^T(t)Px(t)\end{aligned} \tag{2.9}$$

$$V_2(t) = 2\sum_{i=1}^{n} d_i \int_0^{x_i(t)} f_i(s)ds + 2\sum_{i=1}^{n} h_i \int_0^{x_i(t)} (l_i s - f_i(s))ds \tag{2.10}$$

where

$$D = diag(d_1, d_2, \ldots, d_n)(d_i > 0), \quad H = diag(h_1, h_2, \ldots, h_n)(h_i > 0).$$

The derivative of $V_2(t)$ is

$$\dot{V}_2(t) = 2\dot{x}^T(t)Df(x(t)) + 2\dot{x}^T(t)H(Lx(t) - f(x(t))) \tag{2.11}$$

Then we have

$$\begin{aligned}\dot{V}_2(t) + 2\beta V_2(t) = &\ 2\dot{x}^T(t)Df(x(t)) + 2\dot{x}^T(t)HLx(t)\\ &- 2\dot{x}^T(t)Hf(x(t)) + 4\beta x^T(t)Df(x(t))\\ &+ 4\beta x^T(t)HLx(t) - 4\beta x^T(t)Hf(x(t))\end{aligned} \tag{2.12}$$

$$\begin{aligned}V_3(t) = &\int_{t-\tau(t)}^{t} e^{2\beta(s-t)}\begin{bmatrix} x^T(s) & f^T(x(s))\end{bmatrix} Q \begin{bmatrix} x(s) \\ f(x(s))\end{bmatrix} ds\\ &+ \int_{t-\tau}^{t} e^{2\beta(s-t)}\begin{bmatrix} x^T(s) & f^T(x(s))\end{bmatrix} R \begin{bmatrix} x(s) \\ f(x(s))\end{bmatrix} ds\end{aligned} \tag{2.13}$$

where $Q = Q^T = \begin{bmatrix} Q_{11} & Q_{12} \\ * & Q_{22}\end{bmatrix}$, $R = R^T = \begin{bmatrix} R_{11} & R_{12} \\ * & R_{22}\end{bmatrix}$.

The derivative of $V_3(t)$ is

$$\begin{aligned}\dot{V}_3(t) \leq & -2\beta\int_{t-\tau(t)}^{t} e^{2\beta(s-t)}[x^T(s) \quad f^T(x(s))]Q\begin{bmatrix} x(s) \\ f(x(s)) \end{bmatrix}ds \\ & + [x^T(t) \quad f^T(x(t))]Q\begin{bmatrix} x(t) \\ f(x(t)) \end{bmatrix} \\ & - (1-\mu)e^{-2\beta\tau}[x^T(t-\tau(t)) \\ & \quad f^T(x(t-\tau(t)))]Q\begin{bmatrix} x(t-\tau(t)) \\ f(x(t-\tau(t))) \end{bmatrix} \\ & - 2\beta\int_{t-\tau}^{t} e^{2\beta(s-t)}[x^T(s) \quad f^T(x(s))]R\begin{bmatrix} x(s) \\ f(x(s)) \end{bmatrix}ds \\ & + [x^T(t) \quad f^T(x(t))]R\begin{bmatrix} x(t) \\ f(x(t)) \end{bmatrix} \\ & - e^{-2\beta\tau}[x^T(t-\tau) \quad f^T(x(t-\tau))]R\begin{bmatrix} x(t-\tau) \\ f(x(t-\tau)) \end{bmatrix}\end{aligned} \tag{2.14}$$

$$\begin{aligned}\dot{V}_3(t) + 2\beta V_3(t) \leq & [x^T(t) \quad f^T(x(t))]Q\begin{bmatrix} x(t) \\ f(x(t)) \end{bmatrix} \\ & - (1-\mu)e^{-2\beta\tau}[x^T(t-\tau(t)) \quad f^T(x(t-\tau(t)))]Q\begin{bmatrix} x(t-\tau(t)) \\ f(x(t-\tau(t))) \end{bmatrix} \\ & + [x^T(t) \quad f^T(x(t))]R\begin{bmatrix} x(t) \\ f(x(t)) \end{bmatrix} \\ & - e^{-2\beta\tau}[x^T(t-\tau) \quad f^T(x(t-\tau))]R\begin{bmatrix} x(t-\tau) \\ f(x(t-\tau)) \end{bmatrix}\end{aligned} \tag{2.15}$$

$$
\begin{aligned}
V_4(t) = & \int_{t-\tau(t)}^{t} e^{2\beta(s-t)}(x^T(s)LT_1Lx(s) - f^T(x(s))T_1f(x(s)))ds \\
& + \int_{t-\tau}^{t} e^{2\beta(s-t)}(x^T(s)LT_2Lx(s) - f^T(x(s))T_2f(x(s)))ds
\end{aligned}
\tag{2.16}
$$

The derivative of $V_4(t)$ is

$$
\begin{aligned}
\dot{V}_4(t) \leq & -2\beta \int_{t-\tau(t)}^{t} e^{2\beta(s-t)}(x^T(s)LT_1Lx(s) - f^T(x(s))T_1f(x(s)))ds \\
& + x^T(t)LT_1Lx(t) - f^T(x(t))T_1f(x(t)) \\
& - (1-\mu)e^{2\beta\tau}x^T(t-\tau(t))LT_1Lx(t-\tau(t)) \\
& + (1-\mu)e^{2\beta\tau}f^T(x(t-\tau(t)))T_1f(x(t-\tau(t))) \\
& - 2\beta \int_{t-\tau}^{t} e^{2\beta(s-t)}(x^T(s)LT_2Lx(s) - f^T(x(s))T_2f(x(s)))ds \\
& + x^T(t)LT_2Lx(t) - f^T(x(t))T_2f(x(t)) \\
& - e^{2\beta\tau}x^T(t-\tau)LT_2Lx(t-\tau) + e^{2\beta\tau}f^T(x(t-\tau))T_2f(x(t-\tau))
\end{aligned}
\tag{2.17}
$$

$$
\begin{aligned}
\dot{V}_4(t) + 2\beta V_4(t) \leq & \; x^T(t)LT_1Lx(t) - f^T(x(t))T_1f(x(t)) \\
& - (1-\mu)e^{2\beta\tau}x^T(t-\tau(t))LT_1Lx(t-\tau(t)) \\
& + (1-\mu)e^{2\beta\tau}f^T(x(t-\tau(t)))T_1f(x(t-\tau(t))) \\
& + x^T(t)LT_2Lx(t) - f^T(x(t))T_2f(x(t)) \\
& - e^{2\beta\tau}x^T(t-\tau)LT_2Lx(t-\tau) \\
& + e^{2\beta\tau}f^T(x(t-\tau))T_2f(x(t-\tau))
\end{aligned}
\tag{2.18}
$$

From the abovementioned equations, we have $\dot{V}(t) + 2\beta V = \xi^T(t)\Xi\xi(t)$, where $\xi^T(t) = [x^T(t)x^T(t-\tau)x^T(t-\tau(t)) f^T(x(t))f^T(x(t-\tau))f^T(x(t-\tau(t)))]$, which indicates that system (2.3) is a global exponential stable. This completes the proof.

2.1.3 Simulation Example

In this section the example is proposed to show the effectiveness of our results.

Example 2.1: The system of cellular neural networks (CNNs) with time-varying delay has the following parameters (Fig. 2.2):

$$A = \begin{bmatrix} 0.6 & 1 \\ -1 & -1 \end{bmatrix}, \quad B = \begin{bmatrix} 0 & 1.6 \\ -1 & -1 \end{bmatrix}, \quad C = \begin{bmatrix} 1.9 & 0 \\ 0 & 1.4 \end{bmatrix},$$
$$\beta = 0.1, \quad \tau = 1.03, \quad d = 0.1.$$

We can get the matrices through LMI toolbox such that

$$P = \begin{bmatrix} 610.6 & 226.3 \\ 226.3 & 1118.8 \end{bmatrix}, \quad D = \begin{bmatrix} 0.0624 & 0 \\ 0 & 654.781 \end{bmatrix},$$
$$D_1 = \begin{bmatrix} 0.0253 & 0 \\ 0 & 0.0116 \end{bmatrix}, \quad D_2 = \begin{bmatrix} 121.7076 & 0 \\ 0 & 176.6021 \end{bmatrix},$$
$$D_3 = \begin{bmatrix} 0.3318 & 0 \\ 0 & 0.1903 \end{bmatrix}, \quad D_4 = \begin{bmatrix} 686.2998 & 0 \\ 0 & 396.5849 \end{bmatrix},$$

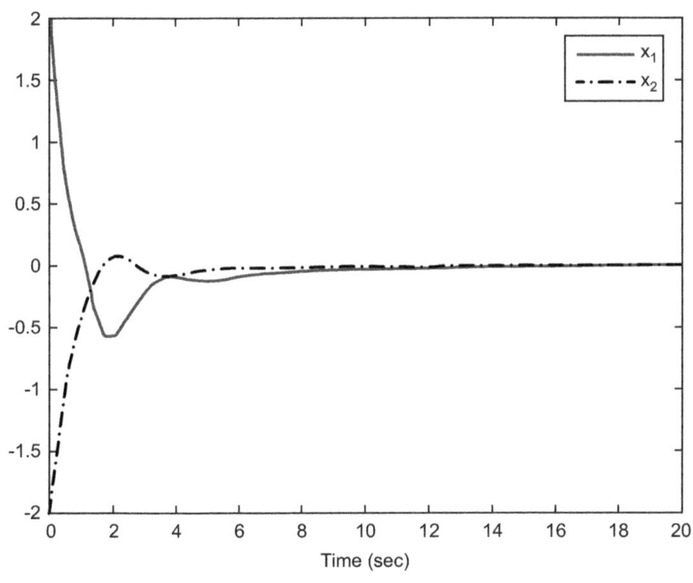

Figure 2.2 The states of the system in Example 2.1.

$$D_5 = \begin{bmatrix} 0.3184 & 0 \\ 0 & 0.929 \end{bmatrix}, \quad D_6 = \begin{bmatrix} 638.4 & 0 \\ 0 & 2651.5 \end{bmatrix},$$

$$T_1 = \begin{bmatrix} 0.0949 & 0 \\ 0 & 0.548 \end{bmatrix}, \quad T_2 = \begin{bmatrix} 480.6708 & 0 \\ 0 & 296.022 \end{bmatrix},$$

$$Q_{11} = \begin{bmatrix} 792.8 & 54.2 \\ 54.2 & 1284.9 \end{bmatrix}, \quad Q_{12} = \begin{bmatrix} 83 & -450.1 \\ 1012.8 & 1073.7 \end{bmatrix},$$

$$Q_{22} = \begin{bmatrix} 800.4 & 821.1 \\ 821.1 & 1207.7 \end{bmatrix},$$

$$R_{11} = \begin{bmatrix} 428.4007 & 147.7588 \\ 147.7588 & 228.6268 \end{bmatrix}, \quad R_{12} = \begin{bmatrix} -0.2467 & -1.2406 \\ 0.0767 & 0.3251 \end{bmatrix},$$

$$R_{22} = \begin{bmatrix} 480.7485 & 0.2205 \\ 0.2205 & 297.0954 \end{bmatrix}.$$

2.2 ROBUST LYAPUNOV STABILITY ANALYSIS OF THE CELLULAR NEURAL NETWORKS BASED ON LINEAR MATRIX INEQUALITY

Neural networks have been applied to a wide range of areas such as associative memory, static image processing, and pattern recognition. In such applications a key factor is to determine the equilibrium and the stability properties of the equilibrium point of the designed neural network. Specially, if a neural network is aimed to be applied to a problem of parallel computation, neural control, and optimization, it must have a unique equilibrium that is globally asymptotically stable. Therefore, stability analysis is very important for the design and applications of neural networks (Arik, 2014). There have been lots of results about the stability conditions of neural networks. A kind of neural network called the cellular neural network (CNN) was proposed and its time delay was considered (Cao and Wang, 2005). There are many results of the stability analysis of CNN, but most of them are delay-independent. However, some delay-dependent results have been proposed (Singh, 2008; Zhang and Yu, 2012; Zhang et al., 2009).

Time delay and uncertainties always occur in system model, and modelling with delay and uncertainties is necessary, because they would cause poor performance or instability. Therefore, the stability of neural networks with uncertain parameters and time delay is very important (Deyou et al., 2007; Wu, 2012).

Although there are many methods to study the stability of neural networks with time-varying delay, there are still some drawbacks that need to be improved. On one hand, conservativeness can be reduced only by constructing new LKFs and declining the enlargement of the derivative of LKF. The LKF usually contains the term $\int_\alpha^\beta \dot{x}^T(s)R\dot{x}(s)ds$. By using a new kind of inequality called the free-matrix-based inequality (Seuret and Gouaisbaut, 2013; He et al., 2016) to deal with $\int_\alpha^\beta \dot{x}^T(s)R\dot{x}(s)ds$, the derived criteria show less conservatism. On the other hand, in all the abovementioned results about neural networks' stability analysis, the free-matrix-based inequality has not been used in the uncertain neural networks, but the uncertain parameters exist in the practical systems. This is the motivation to propose the new study on the robust stability of neural networks with uncertain parameters and time-varying delay. The main contribution of this chapter is that the robust stability condition can be obtained by the theorem presented here, which shows that the method in this literature is less conservative. Also proposed is a new robust stability criterion using the free-matrix-based integral inequality, and as far as we know the free-matrix-based integral inequality has not been used in neural networks with uncertain parameters.

The general form of CNN is

$$\dot{x}(t) = -Cx(t) + Ag(x(t)) + Bg(x(t - d(t))) + S \tag{2.19}$$

where $x(\cdot) = [x_1(\cdot), x_2(\cdot), \ldots, x_n(\cdot)]^T$ is the state vector; $g(x(\cdot)) = [g_1(x_1(\cdot)), g_2(x_2(\cdot)), \ldots, g_n(x_n(\cdot))]^T$ denotes the neuron activation function; $S = [s_1, s_2, \ldots, s_n]^T$ is a constant input vector; C is a diagonal matrix, with $C = diag[c_1 \quad c_2 \quad \ldots \quad c_n]$, $c_i > 0$, $1 \le i \le n$; A and B are the connection weight matrix and the delayed connection weight matrix, respectively; and $d(t)$ is the time-varying delay.

2.2.1 System Formulation and Preliminaries

Consider the following neural network with uncertain parameters and time delay:

$$\dot{z}(t) = -Cz(t) + (A + \Delta A)g(z(t)) + (B + \Delta B)g(z(t - d(t))) + S \tag{2.20}$$

where $z(\cdot) = [z_1(\cdot), z_2(\cdot), \ldots, z_n(\cdot)]^T$ is the state vector; $g(z(\cdot)) = [g_1(z_1(\cdot)), g_2(z_2(\cdot)), \ldots, g_n(z_n(\cdot))]^T$ denotes the neuron activation function; $S = [s_1, s_2, \ldots, s_n]^T$ is a constant input vector, C is a diagonal matrix, with $C = diag[c_1 \quad c_2 \quad \ldots \quad c_n]$, $c_i > 0$, $1 \le i \le n$; A and B are the connection weight matrix and the delayed connection weight matrix, respectively; ΔA and ΔB are uncertain parameters; and $d(t)$ is the time-varying delay.

Assumption 1: Assume that ΔA and ΔB satisfy the following condition:

$$\Delta A = H_1 F_1 E_1, \quad \Delta B = H_2 F_2 E_2 \tag{2.21}$$

Where F_1 and F_2 are unknown constant matrices and satisfy the following condition:

$$F_1^T F_1 \le I, \quad F_2^T F_2 \le I \tag{2.22}$$

Assumption 2: Assume that neuron activation function and the time-varying delay satisfy the following condition:

1. $g(0) = 0, \quad |g(\xi)| \le M$
2. $0 \le g(x) - g(y) \le l_i(x - y), \forall x, y \in R, L = diag(l_1, l_2, \ldots, l_n)$
3. $0 \le \tau(t) \le \tau, \quad \dot{\tau}(t) \le \mu < 1$

Assuming that there is an equilibrium point x^* for the neural networks system (2.20), we have

$$\begin{aligned} \dot{x}^*(t) = &- Cx^*(t) + (A + \Delta A)f\left(x^*(t)\right) + (B + \Delta B) \\ & f\left(x^*(t - \tau(t))\right) + S \end{aligned} \tag{2.23}$$

By the transformation $x(\cdot) = z(\cdot) - z^*$, the equilibrium point can be shifted to the origin, and consequently, system (2.20) can be transferred as

$$\dot{x}(t) = -Cx(t) + \overline{A}f(x(t)) + \overline{B}f(x(t - \tau(t))) \tag{2.24}$$

where $\overline{A} = A + \Delta A, \overline{B} = B + \Delta B$ $x(\cdot) = [x_1(\cdot), x_2(\cdot), \ldots x_n(\cdot)]^T$

$$f(x(\cdot)) = [f_1(x(\cdot)), f_2(x(\cdot)), \ldots f_n(x(\cdot))]^T,$$
$$f_i(x_i(\cdot)) = g_i\left(x_i(\cdot) + z_i^*\right) - g_i\left(z_i^*\right).$$

The function $f_i(x_i(\cdot))$ satisfies that

$$f_i^2(x_i(\cdot)) \leq L_i x_i(\cdot) f_i(x_i(\cdot)) \tag{2.25}$$

where $f_i(0) = 0, i = 1, 2, \ldots n$.

Lemma 1: (Zeng et al., 2015). For any matrix $R > 0$, differentiable signal x in $[\alpha, \beta] \rightarrow \Re^n$. The following inequality holds:

$$\int_\alpha^\beta \dot{x}^T(s) R \dot{x}(s) ds \leq \frac{1}{\beta - \alpha} \varpi^T \widehat{\Omega} \varpi \tag{2.26}$$

where

$$\widehat{\Omega} = \begin{bmatrix} -4R & -2R & 6R \\ * & -4R & 6R \\ * & * & -12R \end{bmatrix},$$

$$\varpi = \begin{bmatrix} x^T(\beta) & x^T(\alpha) & \frac{1}{\beta-\alpha} \int_\alpha^\beta x^T(s) ds \end{bmatrix}^T.$$

Lemma 2: (Zhang and Yu, 2012). Let X and Y be real matrices of appropriate dimensions, and for any scalar $\varepsilon > 0$, the following inequality holds:

$$XY + Y^T X^T \leq \varepsilon^{-1} X X^T + \varepsilon Y^T Y \tag{2.27}$$

Let E, H, and $F(t)$ be real matrices of appropriate dimensions, with $F(t)$ satisfying

$$F^T(t) F(t) \leq I \tag{2.28}$$

Then for any $\varepsilon > 0$ the following inequality holds:

$$EF(t)H + H^T F^T(t)E \leq \varepsilon^{-1}EE^T + \varepsilon H^T H \tag{2.29}$$

2.2.2 Main Results

Theorem 2.2: System (2.24) is global asymptotically stable if there exist the symmetrical positive matrices P, Q, Z and the positive matrices D, D_1, D_2, D_3, D_4, N_1, and L, where L is determined by the active function, and positive scalars ε_i, $i = 1, 2$, such that the following LMI $\Xi < 0$ holds:

$$\Xi = \begin{bmatrix} \Xi_{11} & \Xi_{12} & \Xi_{13} & \Xi_{14} & \Xi_{15} & \Xi_{16} & \Xi_{17} & \Xi_{18} \\ * & \Xi_{22} & 0 & \Xi_{24} & \Xi_{25} & 0 & 0 & 0 \\ * & * & \Xi_{33} & 0 & 0 & \Xi_{36} & \Xi_{37} & \Xi_{38} \\ * & * & * & \Xi_{44} & 0 & \Xi_{46} & 0 & 0 \\ * & * & * & * & \Xi_{55} & 0 & 0 & 0 \\ * & * & * & * & * & \Xi_{66} & \Xi_{67} & \Xi_{68} \\ * & * & * & * & * & * & \Xi_{77} & 0 \\ * & * & * & * & * & * & * & \Xi_{88} \end{bmatrix} \tag{2.30}$$

where

$$\Xi_{11} = -PC - CP + Q_{11} + 2LD_1 - 4\tau^{-1}N_1,$$
$$\Xi_{12} = -2\tau^{-1}N_1,$$
$$\Xi_{13} = PA + Q_{12} - CD - D_1 + LD_2,$$
$$\Xi_{14} = PB,$$
$$\Xi_{15} = 6\tau^{-1}N_1,$$
$$\Xi_{16} = -\tau C^T Z,$$
$$\Xi_{17} = PH_1,$$
$$\Xi_{18} = PH_2,$$
$$\Xi_{22} = 2LD_3 + (\mu - 1)Q_{11} - 4\tau^{-1}N_1,$$
$$\Xi_{24} = LD_4 - D_3 + (\mu - 1)Q_{12},$$
$$\Xi_{25} = 6\tau^{-1}N_1,$$
$$\Xi_{33} = Q_{22} + DA + A^T D - 2D_2 + \varepsilon_1 E_1^T E_1 \Xi_{34} + DB,$$

$$\begin{aligned}
\Xi_{36} &= -\tau A^T Z, \\
\Xi_{37} &= DH_1, \\
\Xi_{38} &= DH_2, \\
\Xi_{44} &= -2D_4 + (\mu - 1)Q_{22} + \varepsilon_2 E_2^T E_2, \\
\Xi_{46} &= -\tau B^T Z, \\
\Xi_{55} &= -12\tau^{-1} N_1, \\
\Xi_{66} &= -\tau Z, \\
\Xi_{67} &= \tau Z H_1, \\
\Xi_{68} &= \tau Z H_2, \\
\Xi_{77} &= -\varepsilon_1 I, \\
\Xi_{88} &= -\varepsilon_2 I.
\end{aligned}$$

Proof: Choose the following LKF candidate:

$$V(t) = V_1(t) + V_2(t) + V_3(t) + V_4(t) \tag{2.31}$$

$$V_1(t) = x^T(t) P x(t) \tag{2.32}$$

where $P = P^T > 0$

$$V_2(t) = \int_{t-\tau(t)}^{t} (\tau + s - t)\dot{x}^T(s) Z \dot{x}(s) ds \tag{2.33}$$

where $Z = Z^T > 0$

$$V_3(t) = 2\sum_{i=1}^{n} d_i \int_0^{x_i(t)} f_i(s) ds \tag{2.34}$$

where $D = diag(d_1, d_2, \ldots d_n), (d_i > 0)$

$$V_4(t) = \int_{t-\tau(t)}^{t} \begin{bmatrix} x(s) & f(x(s)) \end{bmatrix}^T Q \begin{bmatrix} x(s) \\ f(x(s)) \end{bmatrix} ds \tag{2.35}$$

where $Q = Q^T = \begin{pmatrix} Q_{11} & Q_{12} \\ Q_{21} & Q_{22} \end{pmatrix}$.

The derivative of $V(t)$ is

$$\begin{aligned}
\dot{V}_1(t) &= 2x^T(t) P \dot{x}(t) \\
&= 2x^T(t) P\left[- Cx(t) + \overline{A} f(x(t)) + \overline{B} f(x(t - \tau(t)))\right]
\end{aligned} \tag{2.36}$$

$$\dot{V}_2(t) = \tau \dot{x}^T(t) Z \dot{x}(t) - \int_{t-\tau(t)}^{t} \dot{x}^T(s) Z \dot{x}(s) ds \tag{2.37}$$

According to Lemma 1, we have

$$\int_{t-\tau(t)}^{t} \dot{x}^T(s) Z \dot{x}(s) ds \leq \frac{1}{\tau(t)} \begin{bmatrix} e_1^T & e_2^T & e_5^T \end{bmatrix}^T \begin{bmatrix} -4N_1 & -2N_1 & 6N_1 \\ * & -4N_1 & 6N_1 \\ * & * & -12N_1 \end{bmatrix} [\begin{bmatrix} e_1^T & e_2^T & e_5^T \end{bmatrix}] \tag{2.38}$$

It is easy to see that

$$\begin{aligned} \dot{V}_2(t) &\leq \tau \dot{x}^T(t) Z \dot{x}(t) \\ &\leq \tau\big(- Cx(t) + \overline{A} f(x(t)) + \overline{B} f(x(t-\tau(t)))\big)^T \\ &\quad Z\big(- Cx(t) + \overline{A} f(x(t)) + \overline{B} f(x(t-\tau(t)))\big) \\ &\quad - \frac{1}{\tau(t)} \begin{bmatrix} e_1^T & e_2^T & e_5^T \end{bmatrix}^T \begin{bmatrix} -4N_1 & -2N_1 & 6N_1 \\ * & -4N_1 & 6N_1 \\ * & * & -12N_1 \end{bmatrix} \begin{bmatrix} e_1^T & e_2^T & e_5^T \end{bmatrix} \end{aligned} \tag{2.39}$$

$$\begin{aligned} \dot{V}_3(t) &= 2 \sum_{i=1}^{n} d_i \dot{x}_i(t) f_i(x_i(t)) \\ &= -2 f^T(x(t)) D C x(t) + 2 f^T(x(t)) D \overline{A} f(x(t)) \\ &\quad + 2 f^T(x(t)) D \overline{B} f(x(t-\tau(t))) \end{aligned} \tag{2.40}$$

$$\dot{V}_4(t) \leq \begin{bmatrix} e_1^T & e_3^T \end{bmatrix}^T Q \begin{bmatrix} e_1^T \\ e_3^T \end{bmatrix} - (1-\mu) \begin{bmatrix} e_2^T & e_4^T \end{bmatrix}^T Q \begin{bmatrix} e_2^T \\ e_4^T \end{bmatrix} \tag{2.41}$$

For any symmetrical matrix $D_i, i = 1, 2, 3, 4$, we have $\boldsymbol{\Delta}_1 > 0$, $\boldsymbol{\Delta}_2 > 0$, where

$$\begin{aligned}\Delta_1 = {} & 2x^T(t)LD_1x(t) - x^T(t)D_1f(x(t)) - f^T(x(t))D_1x(t) \\ & + x^T(t)LD_2f(x(t)) + f^T(x(t))D_2Lx(t) - 2f^T(x(t))D_2f(x(t))\end{aligned} \tag{2.42}$$

$$\begin{aligned}\Delta_2 = {} & 2x^T(t-\tau(t))LD_3x(t-\tau(t)) \\ & -x^T(t-\tau(t))D_3f(x(t-\tau(t))) \\ & -f^T(x(t-\tau(t)))D_3x(t-\tau(t)) \\ & +x^T(t-\tau(t))LD_4f(x(t-\tau(t))) \\ & +f^T(x(t-\tau(t)))D_4Lx(t-\tau(t)) \\ & -2f^T(x(t-\tau(t)))D_4f(x(t-\tau(t)))\end{aligned} \tag{2.43}$$

Adding (2.36–2.43), we have

$$\dot{V} = \dot{V}_1 + \dot{V}_2 + \dot{V}_3 + \dot{V}_4 + \Delta_1 + \Delta_2 = e^T(t)\Xi_1 e(t) < 0 \tag{2.44}$$

where

$$e^T(t) = \begin{bmatrix} e_1^T(t) & e_2^T(t) & e_3^T(t) & e_4^T(t) & e_5^T(t) \end{bmatrix} \tag{2.45}$$

$$\begin{aligned} e_1^T(t) &= x^T(t), \\ e_2^T(t) &= x^T(t-\tau(t)), \\ e_3^T(t) &= f^T(x(t)), \\ e_4^T(t) &= f^T(x(t-\tau(t))), \\ e_5^T(t) &= \frac{1}{\tau(t)}\int_{\tau(t)}^{t-\tau(t)} x^T(s)ds \end{aligned}$$

$$\Xi_1 = \begin{bmatrix} \Phi_{11} & \Phi_{12} & \Phi_{13} & \Phi_{14} & \Phi_{15} \\ * & \Phi_{22} & 0 & \Phi_{24} & \Phi_{25} \\ * & * & \Phi_{33} & \Phi_{34} & 0 \\ * & * & * & \Phi_{44} & 0 \\ * & * & * & * & \Phi_{55} \end{bmatrix} \tag{2.46}$$

where

$$\Phi_{11} = -PC - CP + Q_{11} - \tau^{-1}Z + 2LD_1 + \tau CZC - 4\tau^{-1}N_1,$$

$$\Phi_{12} = \tau^{-1}Z,$$

$$\Phi_{13} = P\overline{A} + Q_{12} - \tau C^T Z\overline{A} - CD - D_1 + LD_2,$$

$$\Phi_{14} = P\overline{B} - \tau CZ\overline{B},$$
$$\Phi_{15} = 6\tau^{-1}N_1,$$
$$\Phi_{22} = 2LD_3 - \tau^{-1}Z + (\mu - 1)Q_{11},$$
$$\Phi_{24} = LD_4 - D_3 + (\mu - 1)Q_{12},$$
$$\Phi_{25} = 6\tau^{-1}N_1,$$
$$\Phi_{33} = Q_{22} + D\overline{A} + \overline{A}^T D - 2D_2 + \tau\overline{A}^T Z\overline{A},$$
$$\Phi_{34} = D\overline{B} + \tau\overline{A}^T Z\overline{B},$$
$$\Phi_{44} = -2D_4 + (\mu - 1)Q_{22} + \tau\overline{B}^T Z\overline{B},$$
$$\Phi_{55} = -12\tau^{-1}N_1.$$

According to Lemma 2, $\Xi_1 < 0$ is equal to $\Xi_2 < 0$.where

$$\Xi_2 = \begin{bmatrix} \Theta_{11} & \Theta_{12} & \Theta_{13} & \Theta_{14} & \Theta_{15} & \Theta_{16} \\ * & \Theta_{22} & 0 & \Theta_{24} & \Theta_{25} & 0 \\ * & * & \Theta_{33} & \Theta_{34} & 0 & \Theta_{36} \\ * & * & * & \Theta_{44} & 0 & 0 \\ * & * & * & * & \Theta_{55} & 0 \\ * & * & * & * & * & \Theta_{66} \end{bmatrix} \tag{2.47}$$

where

$$\Theta_{11} = -PC - CP + Q_{11} - \tau^{-1}Z + 2LD_1,$$
$$\Theta_{12} = \tau^{-1}Z,$$
$$\Theta_{13} = P\overline{A} + Q_{12} - CD - D_1 + LD_2,$$
$$\Theta_{14} = P\overline{B},$$
$$\Theta_{15} = 6\tau^{-1}N_1,$$
$$\Theta_{16} = -\tau C^T Z,$$
$$\Theta_{22} = 2LD_3 - \tau^{-1}Z + (\mu - 1)Q_{11},$$
$$\Theta_{24} = LD_4 - D_3 + (\mu - 1)Q_{12},$$
$$\Theta_{25} = 6\tau^{-1}N_1,$$
$$\Theta_{33} = Q_{22} + D\overline{A} + \overline{A}^T D - 2D_2,$$
$$\Theta_{34} = D\overline{B},$$

$$\begin{aligned}
\Theta_{36} &= -\tau\overline{A}^T Z,\\
\Theta_{44} &= -2D_4 + (\mu - 1)Q_{22},\\
\Theta_{45} &= -\tau\overline{B}^T Z,\\
\Theta_{55} &= -12\tau^{-1}N_1,\\
\Theta_{66} &= -\tau Z.
\end{aligned}$$

$\Xi_2 < 0$ contains uncertain matrices $\overline{A}, \overline{B}$. Substitute (2.21) into $\Xi_2 < 0$

$$\Xi_2 = \Xi_3 + \Omega_1 F_1 \Sigma_1 + \Sigma_1^T F_1^T \Omega_1^T + \Omega_2 F_2 \Sigma_2 + \Sigma_2^T F_2^T \Omega_2^T \tag{2.48}$$

where

$$\Omega_1 = \begin{bmatrix} PH_1 \\ 0 \\ DH_1 \\ 0 \\ \tau Z H_1 \end{bmatrix}, \quad \Omega_2 = \begin{bmatrix} PH_2 \\ 0 \\ DH_2 \\ 0 \\ \tau Z H_2 \end{bmatrix}.$$

$$\Xi_3 = \begin{bmatrix} \Psi_{11} & \Psi_{12} & \Psi_{13} & \Psi_{14} & \Psi_{15} & \Psi_{16} \\ * & \Psi_{22} & 0 & \Psi_{24} & \Psi_{25} & 0 \\ * & * & \Psi_{33} & \Psi_{34} & 0 & \Psi_{36} \\ * & * & * & \Psi_{44} & 0 & 0 \\ * & * & * & * & \Psi_{55} & 0 \\ * & * & * & * & * & \Psi_{66} \end{bmatrix} \tag{2.49}$$

where

$$\begin{aligned}
\Psi_{11} - PC - CP + Q_{11} - \tau^{-1}Z + 2LD_1,&\\
\Psi_{12} &= \tau^{-1}Z,\\
\Psi_{13} &= PA + Q_{12} - CD - D_1 + LD_2,\\
\Psi_{14} &= PB,\\
\Psi_{15} &= 6\tau^{-1}N_1,\\
\Psi_{16} &= -\tau C^T Z,\\
\Psi_{22} &= 2LD_3 - \tau^{-1}Z + (\mu - 1)Q_{11},\\
\Psi_{24} &= LD_4 - D_3 + (\mu - 1)Q_{12},
\end{aligned}$$

$$\Psi_{25} = 6\tau^{-1}N_1,$$
$$\Psi_{33} = Q_{22} + DA + A^T D - 2D_2,$$
$$\Psi_{34} = DB,$$
$$\Psi_{36} = -\tau A^T Z,$$
$$\Psi_{44} = -2D_4 + (\mu - 1)Q_{22},$$
$$\Psi_{46} = -\tau B^T Z,$$
$$\Psi_{55} = -12\tau^{-1}N_1,$$
$$\Psi_{66} = -\tau Z.$$

According to Lemma 2, we have $\Xi < 0$, that is, $\dot{V}(x(t)) < 0$. So system (2.24) is global asymptotically stable. This completes the proof.

Remark 1: If the parameters $\Delta A = \Delta B = 0$, then system (2.24) should be rewritten as follows:

$$\dot{x}(t) = -Cx(t) + Af(x(t)) + Bf(x(t - \tau(t))) \tag{2.50}$$

Corollary 2.1: System (2.50) is global asymptotically stable if there exist the symmetrical positive matrices P, Q, Z and the positive matrices D, D_1, D_2, D_3, D_4, N_1, and L, where L is determined by the active function, such that the following LMI $\Xi < 0$ holds:

$$\Xi = \begin{bmatrix} \Xi_{11} & \Xi_{12} & \Xi_{13} & \Xi_{14} & \Xi_{15} & \Xi_{16} \\ * & \Xi_{22} & 0 & \Xi_{24} & \Xi_{25} & 0 \\ * & * & \Xi_{33} & 0 & 0 & \Xi_{36} \\ * & * & * & \Xi_{44} & 0 & \Xi_{46} \\ * & * & * & * & \Xi_{55} & 0 \\ * & * & * & * & * & \Xi_{66} \end{bmatrix} \tag{2.51}$$

where

$$\Xi_{11} = -PC - CP + Q_{11} + 2LD_1 - 4\tau^{-1}N_1,$$
$$\Xi_{12} = -2\tau^{-1}N_1,$$
$$\Xi_{13} = PA + Q_{12} - CD - D_1 + LD_2,$$
$$\Xi_{14} = PB,$$

$$\Xi_{15} = 6\tau^{-1}N_1,$$
$$\Xi_{16} = -\tau C^T Z,$$
$$\Xi_{22} = 2LD_3 + (\mu - 1)Q_{11} - 4\tau^{-1}N_1,$$
$$\Xi_{24} = LD_4 - D_3 + (\mu - 1)Q_{12},$$
$$\Xi_{25} = 6\tau^{-1}N_1,$$
$$\Xi_{33} = Q_{22} + DA + A^T D - 2D_2 + \varepsilon_1 E_1^T E_1 \Xi_{34} + DB,$$
$$\Xi_{36} = -\tau A^T Z,$$
$$\Xi_{44} = -2D_4 + (\mu - 1)Q_{22} + \varepsilon_2 E_2^T E_2,$$
$$\Xi_{46} = -\tau B^T Z,$$
$$\Xi_{55} = -12\tau^{-1}N_1,$$
$$\Xi_{66} = -\tau Z.$$

The proof process is similar to that of Theorem 2.2.

2.2.3 Simulation Example

Three numerical examples are given in this section to illustrate the effectiveness of our method proposed in Theorem 2.2.

Example 2.2: Consider system (2.24) with the following parameters:

$$C = \begin{pmatrix} 5 & 0 \\ 0 & 4 \end{pmatrix}, \quad A = \begin{pmatrix} -1.25 & 0.5 \\ -0.5 & 0.75 \end{pmatrix}, \quad B = \begin{pmatrix} 0.75 & 0.5 \\ -0.5 & 0.75 \end{pmatrix},$$

$$H_1 = H_2 = E_1 = E_2 = \begin{pmatrix} 0.5 & 0 \\ 0 & 0.5 \end{pmatrix},$$

$$F_1 = F_2 = \begin{pmatrix} 1 & 0 \\ 0 & 1 \end{pmatrix}, \quad f(x) = \arctan x, \quad \tau(t) = 2.1 + 0.9 \sin t.$$

According to Theorem 2.2, we can get

$$D = \begin{pmatrix} 2.9454 & 0 \\ 0 & 3.3471 \end{pmatrix}, \quad D_1 = \begin{pmatrix} 3.1030 & 0 \\ 0 & 4.6289 \end{pmatrix},$$

$$D_2 = \begin{pmatrix} 9.7798 & 0 \\ 0 & 10.3986 \end{pmatrix}, \quad D_3 = \begin{pmatrix} 0.5773 & 0 \\ 0 & 0.7246 \end{pmatrix},$$

$$D_4 = \begin{pmatrix} 3.0926 & 0 \\ 0 & 1.8786 \end{pmatrix}, \quad P = \begin{pmatrix} 4.4464 & -0.6934 \\ -0.6934 & 4.4426 \end{pmatrix},$$

$$Z = \begin{pmatrix} 0.0088 & -0.0037 \\ -0.0037 & 0.0051 \end{pmatrix}, \quad N_1 = \begin{pmatrix} 2.8388 & 1.0763 \\ 0.9636 & 6.7183 \end{pmatrix},$$

$$Q_{11} = \begin{pmatrix} 18.9296 & -4.2831 \\ -4.2831 & 10.5957 \end{pmatrix}, \quad Q_{12} = \begin{pmatrix} 9.8365 & -1.7228 \\ -0.8106 & 0.9329 \end{pmatrix},$$

$$Q_{22} = \begin{pmatrix} 15.0379 & -0.4572 \\ -0.4572 & 6.2812 \end{pmatrix},$$

$$\varepsilon_1 = 7.7614, \quad \varepsilon_2 = 3.5858.$$

This example is presented in the work by Deyou et al. (2007). In their work the maximum admissible upper bound of τ is 3 and the maximum admissible upper bound of μ is 0.8. The robust stability condition cannot be obtained in their work when the upper bound of μ is changed to 0.9. However, the system is stable. But in this chapter, Theorem 2.2 can obtain the robust stability condition when the upper bound of μ is changed to 0.9, which shows that their method is more conservative than ours (Fig. 2.3).

Example 2.3: Consider system (2.50) with the following parameters (Fig. 2.4):

$$C = \begin{pmatrix} 5 & 0 \\ 0 & 4 \end{pmatrix}, \quad A = \begin{pmatrix} -1 & 0.5 \\ -0.5 & 1 \end{pmatrix}, \quad B = \begin{pmatrix} 1 & 0.5 \\ -0.5 & 1 \end{pmatrix},$$

$$f(x) = \arctan x, \quad \tau(t) = 2.1 + 0.9 \sin t.$$

According to Corollary 1, we can get

$$D = \begin{pmatrix} 3.6081 & 0 \\ 0 & 3.6805 \end{pmatrix}, \quad D_1 = \begin{pmatrix} 3.9046 & 0 \\ 0 & 5.3228 \end{pmatrix},$$

$$D_2 = \begin{pmatrix} 11.1041 & 0 \\ 0 & 11.4827 \end{pmatrix}, \quad D_3 = \begin{pmatrix} 0.7202 & 0 \\ 0 & 0.7949 \end{pmatrix},$$

$$D_4 = \begin{pmatrix} 3.5268 & 0 \\ 0 & 2.0215 \end{pmatrix}, \quad P = \begin{pmatrix} 5.3355 & -0.6862 \\ -0.6862 & 4.8839 \end{pmatrix},$$

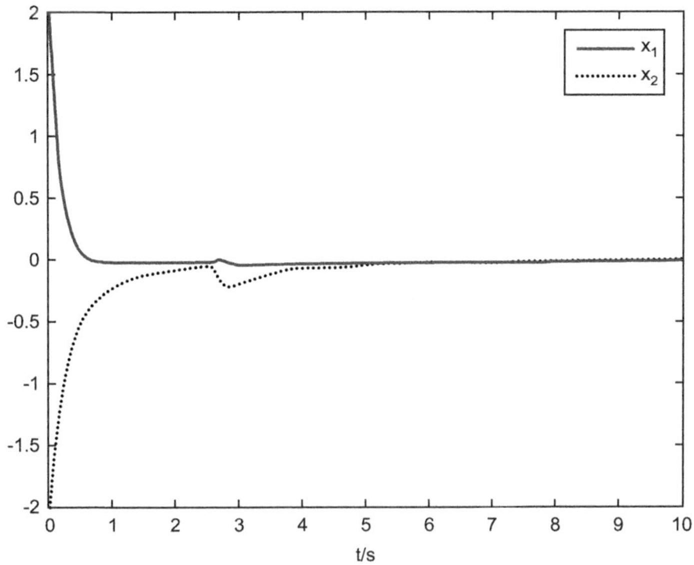

Figure 2.3 The states of the system in Example 2.2 with $\tau(t) = 2.1 + 0.9 \sin t$.

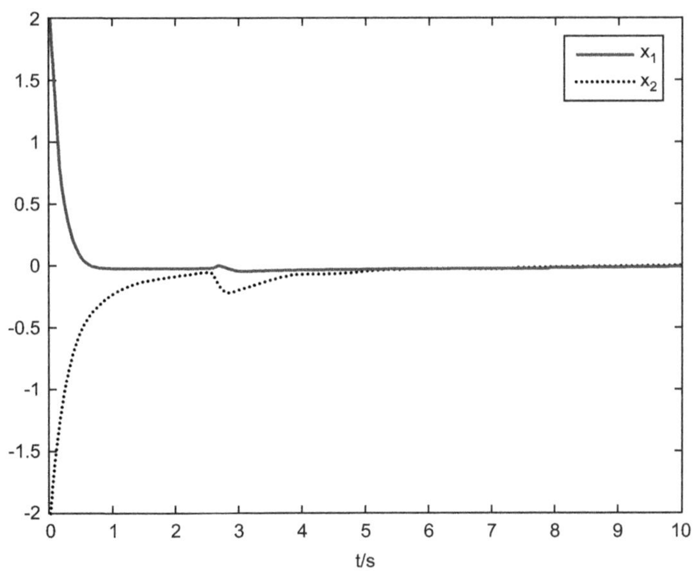

Figure 2.4 The states of the system in Example 2.3 with $\tau(t) = 2.1 + 0.9 \sin t$.

$$Z = \begin{pmatrix} 0.0094 & -0.0032 \\ -0.0032 & 0.0040 \end{pmatrix}, \quad N_1 = \begin{pmatrix} 3.3644 & 1.1567 \\ 1.0472 & 7.5419 \end{pmatrix},$$

$$Q_{11} = \begin{pmatrix} 22.5553 & -4.3646 \\ -4.3646 & 11.4618 \end{pmatrix}, \quad Q_{12} = \begin{pmatrix} 11.5192 & -2.0121 \\ -0.5728 & 0.3975 \end{pmatrix},$$

$$Q_{22} = \begin{pmatrix} 17.9299 & -0.6993 \\ -0.6993 & 7.1739 \end{pmatrix}.$$

Remark 2: If the time delay of Example 2.3 changes to $\tau(t) = 10 + 10 \sin t$, the system is not stable. There are two reasons for this, first, when $t = 0$ or $t = \pi$ the time delay is $\tau(t) = 0$, but in this chapter, we need $\tau(t) > 0$ and second, the derivative of time delay $\dot{\tau}(t) \leq 20$, but in this chapter, we need $\dot{\tau}(t) < 1$. That is, the stability criterion in this chapter is delay dependent (Fig. 2.5).

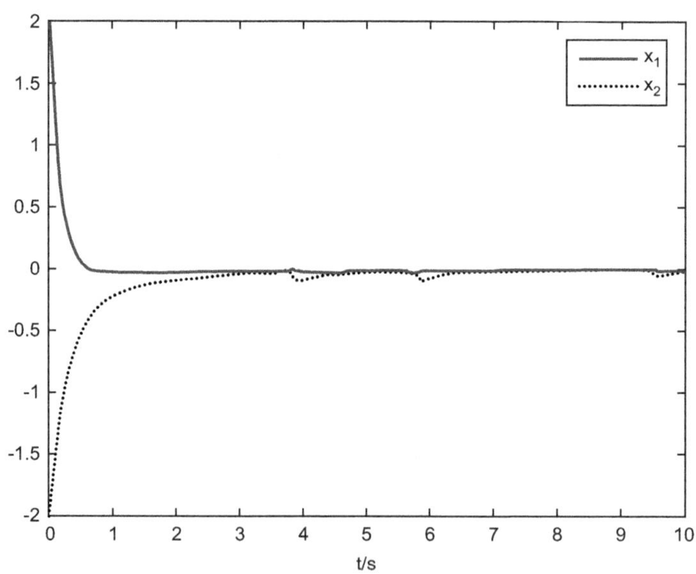

Figure 2.5 The states of the system in Example 2.3 with $\tau(t) = 10 + 10 \sin t$.

Example 2.4: Consider system (2.24) with the following parameters:

$$C = \begin{pmatrix} 0.7 & 0 \\ 0 & 0.9 \end{pmatrix}, \quad A = \begin{pmatrix} 0.6 & 1 \\ -1 & -1 \end{pmatrix}, \quad B = \begin{pmatrix} 0 & 1.6 \\ 1 & 1 \end{pmatrix},$$

$$H_1 = H_2 = E_1 = E_2 = \begin{pmatrix} 0.5 & 0 \\ 0 & 0.5 \end{pmatrix}, F_1 = F_2 = \begin{pmatrix} 1 & 0 \\ 0 & 1 \end{pmatrix},$$

$$f(x) = \arctan x.$$

The system is delay dependent as follows. The system is stable only when the time delay is very small (Figs. 2.6 and 2.7).

This simulation indicates that the system stability can be proved with Theorem 2.2 if the delay is 0.0962, whereas in the work by Deyou et al. (2007) the system stability can be proved if the delay is 0.0135. This means that the results in this chapter are better.

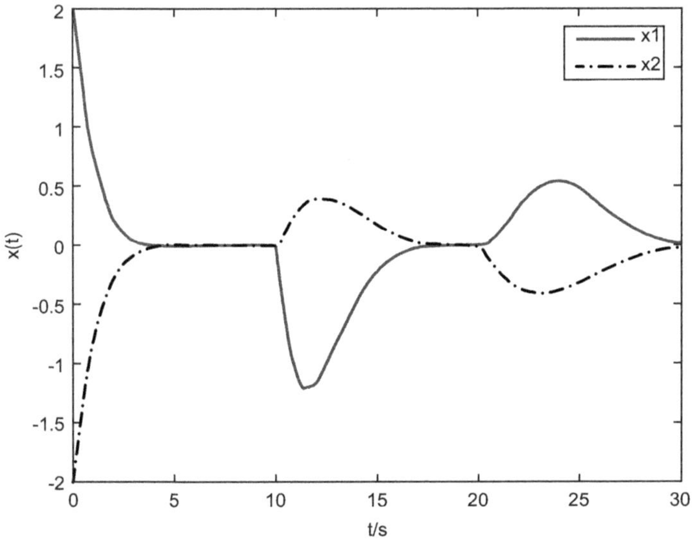

Figure 2.6 The states of the system in Example 2.4 with $\tau = 20$.

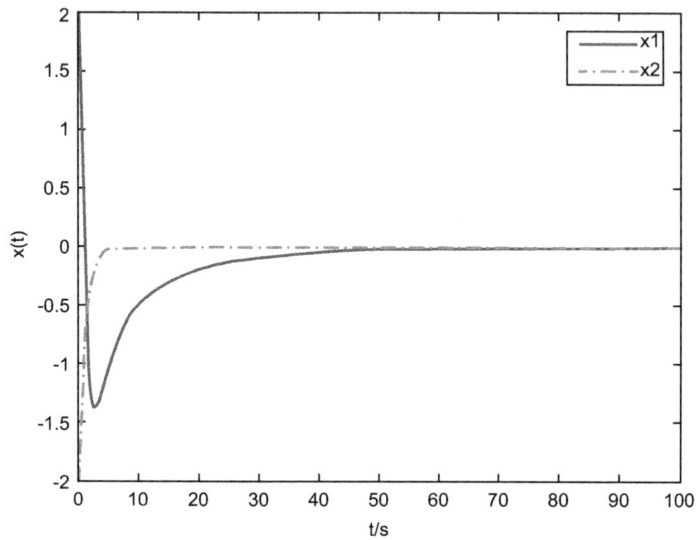

Figure 2.7 The states of the system in Example 2.4 with $\tau = 0.0962$.

Example 2.5: Consider the ball and plate system (Kassem et al., 2015):

$$\begin{cases} \left(m_b + \frac{I_b}{r_b^2}\right)\ddot{x} + m_b g\alpha = 0 \\ \left(m_b + \frac{I_b}{r_b^2}\right)\ddot{y} + m_b g\beta = 0 \end{cases}$$

with neural network identification method (Zhang et al., 2008) to design the controller. The system is stable if the parameters of neural networks satisfy the theorem by the identification process or design the controller to satisfy the theorem. For the neural networks the parameter C plays an important role in stability, so it is easy to design the stable neural networks through a parameter.

The ball and plate system can be controlled as in the following (Figs. 2.8 and 2.9).

The simulations indicate that the stability result can be used to control the ball and plate system.

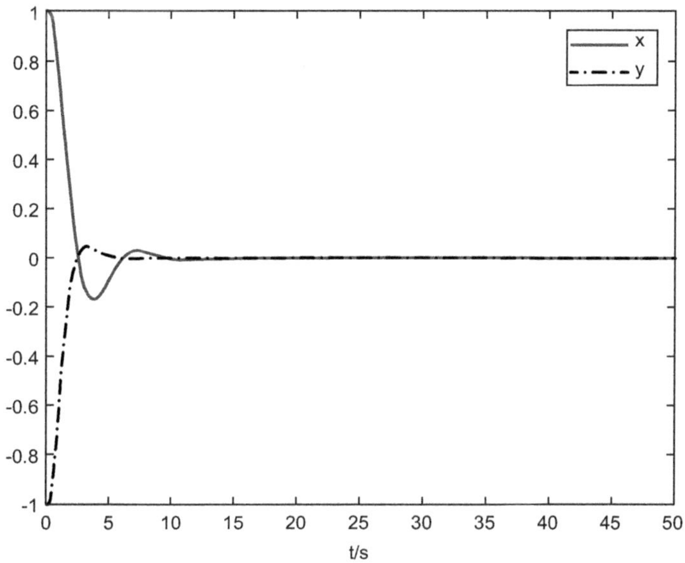

Figure 2.8 The states of the ball and plate system.

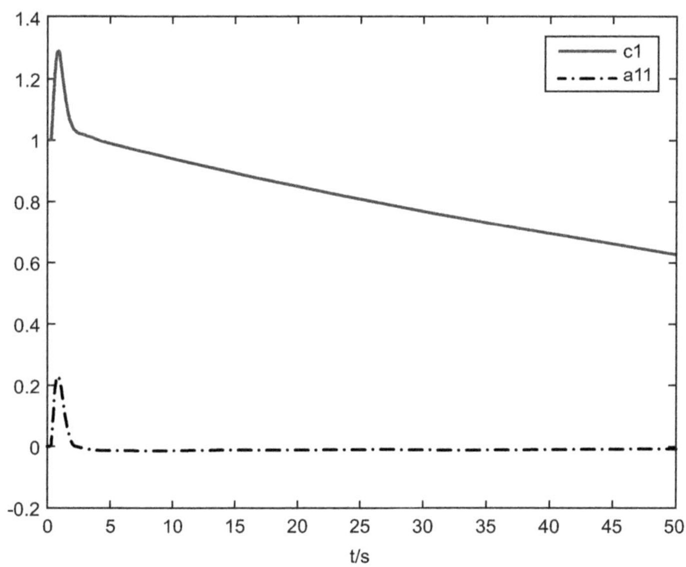

Figure 2.9 The parameters of the identification neural networks.

REFERENCES

Arik, S., 2014. New criteria for global robust stability of delayed neural networks with norm-bounded uncertainties. IEEE Transactions on Neural Networks and Learning Systems 25, 1045–1052.

Cao, J., Wang, J., 2005. Global asymptotic and robust stability of recurrent neural networks with time delays. Circuits and Systems I Regular Papers IEEE Transactions on 52, 417–426.

Deyou, L., Jianhua, Z., Xinping, G., 2007. Exponential stability for cellular neural networks: an LMI approach. Journal of Systems Engineering and Electronics 18, 68–71.

Haoran He, W., Jun Yang, Z., Na, J., Gao, G., 2018. Unknown dynamics estimator-based control for a 3-DOF model helicopter system. International Journal of Applied Mathematics in Control Engineering 1, 62–69.

He, Y., Ji, M.-D., Zhang, C.-K., Wu, M., 2016. Global exponential stability of neural networks with time-varying delay based on free-matrix-based integral inequality. Neural Networks 77, 80–86.

Kassem, A., Haddad, H., Albitar, C., 2015. Comparison between different methods of control of ball and plate system with 6DOF stewart platform. IFAC-Papers OnLine 48, 47–52.

Seuret, A., Gouaisbaut, F., 2013. Wirtinger-based integral inequality: application to time-delay systems ☆. Automatica 49, 2860–2866.

Singh, V., 2008. Improved global robust stability of interval delayed neural networks via split interval: Generalizations. Applied Mathematics and Computation 206, 290–297.

Wu, W.J., 2018. Parametric Lyapunov equation approach to stabilization and set invariance conditions of linear systems with input saturation. International Journal of Applied Mathematics in Control Engineering 1, 79–84.

Wu, X.L., 2012. Synchronisation of unified chaotic systems with uncertain parameters in finite time. International Journal of Modelling, Identification and Control 17, 295–301.

Zeng, H., He, Y., Wu, M., She, J., 2015. Free-matrix-based integral inequality for stability analysis of systems with time-varying delay. IEEE Transactions on Automatic Control 60, 1.

Zhang, D., Yu, L., 2012. Exponential stability analysis for neutral switched systems with interval time-varying mixed delays and nonlinear perturbations. Nonlinear Analysis Hybrid Systems 6, 775–786.

Zhang, J., Li, H., Guan, X., Wu, X., 2008. Delay-dependent asymptotic stability of cellular neural networks with time varying delay. In: 2008 7th World Congress on Intelligent Control and Automation, pp. 2890–2893.

Zhang, Q., Wei, X., Xu, J., 2009. Delay-dependent exponential stability of cellular neural networks with time-varying delays. Chaos Solitons Fractals. Chaos, Solitons & Fractals 23, 1363–1369.

CHAPTER 3

Classical Nonlinear Systems Control Methods

Contents

Chapter 3 studies classical nonlinear system control methods, such as feedback control and backstepping control. First, for a class of nonlinear system, feedback controller is designed based on Lyapunov stability theory, to achieve asymptotic stability of the closed-loop systems. Second, backstepping controller is designed, based on Lyapunov stability theory, to make output of the system achieve the given reference signal (Fig. 3.1).

3.1 SLIDING MODE CONTROL

It has been well understood that the effect of uncertainties may destroy or corrupt many man-made (such as industrial) automatic control systems (Lee et al., 2012). For such reason, research on uncertainties for dynamical systems has been well attended as one of the important areas to satisfy the increasing demand for reliable practical control system operation (Marino and Tomei, 1993; Xiaoyuan Luo, 2018).

Adaptive Sliding Mode Neural Network Control for Nonlinear Systems
ISBN 978-0-12-815372-7
https://doi.org/10.1016/B978-0-12-815372-7.00003-3

Figure 3.1 The structure chart of Chapter 3.

The variable structure theory, the foundation of sliding mode control, was proposed by Emelynaov, Taran, and Utkin in 1950s. This is a kind of phase plane—based design method accommodated with strong robustness. A variable structure system can be viewed as a one composed of independent structures together with a switching function between each of the structures (Żak, 2003). In general, there are two classes of variable structure systems: the variable structure systems with sliding mode and the variable structure systems without sliding mode (such as variable structure PID control system). Itkis' monograph (Itkis, 1976) and Utkin's famous article (Utkin, 1977) established the fundamental framework for sliding mode control theory. It should be noted that the original sliding mode control theory is more complicated until middle of 1980s.

Sliding mode control theory has experienced three stages of development: the first stage has been on the research of sliding mode control of single input and single output linear systems by using error and its derivative as state variables. The second stage has been on the research of sliding mode control of multi input and multi output linear and nonlinear systems since 1960s. The third stage has been research of sliding mode control of discrete-time systems, time-delay systems, distributed parameter system, and other complex systems since 1980s. At the same time, adaptive, neural network, fuzzy, genetic, and other mechanisms also have been introduced to upgrade the functionality and performance in sliding mode control system design.

Generally, the design of sliding mode controllers consists of two steps (Moreno and Osorio, 2012) (Wu, 2012): (1) construction of sliding surface to achieve the desired system behavior; and (2) selecting controller to stable on the sliding surface.

Consequently, there have been a number of sliding mode approaches proposed for both continuous-time and discrete-time sliding mode controllers. For example, Sarpturk et al. (1987), Wu et al. (2014), and Li et al. (2014) proposed a sliding and convergence condition for controlling discrete-time systems, Cheng et al. (2000) proposed a simple methodology for designing sliding mode controllers for a class of linear multiinput discrete-time systems with matching perturbations. Moreno and Osorio (2012) and Zhang et al. (2017) proposed a method to construct a family of strict Lyapunov functions, for the super-twisting algorithm, with or without perturbations.

The main results include description of plant model, definition of U-model, and design of U-model—enabled super-twisting control systems.

3.1.1 Design of Supertwisting Control Systems

First of all, a discrete-time supertwisting system is construct as following,

$$\begin{aligned} x_1(k+1) &= \rho_1 x_1(k) + \tau x_2(k) - \tau k_1 |x_1(k)|^{\frac{1}{2}} sign(x_1(k)) \\ x_2(k+1) &= \rho_2 x_2(k) - \tau k_2 sign(x_1(k)) \end{aligned} \tag{3.1}$$

where $x_i(k) \in \Re$ are the states, $\rho_i \in \Re^+$, $\rho_i < 1$, $k_i \in \Re$ for $i = 1,2$ are the gains to be designed for ensuring the convergence of (3.1) to the origin, and $\tau > 0$ is the sampling period.

The sign function $sign(x)$ is defined as

$$sign(x) := \begin{cases} -1 & \text{if} \quad x < 0 \\ 0 & \text{if} \quad x = 0 \\ 1 & \text{if} \quad x > 0 \end{cases} \tag{3.2}$$

where $x \in \Re$.

Thus, the system can be represented as

$$x(k+1) = Ax(k) + B(k) sign(x_1(k)) \tag{3.3}$$

where

$$A = \begin{bmatrix} \rho_1 & \tau \\ 0 & \rho_2 \end{bmatrix} \quad B(k) = \begin{bmatrix} -\tau k_1 |x_1(k)|^{\frac{1}{2}} \\ -\tau k_2 \end{bmatrix} \tag{3.4}$$

Remark 1: The trajectories of the supertwisting reference system converge asymptotically to a ball while certain conditions are satisfied.

Remark 2: The constructed supertwisting system above can be used as desired reference for controlling a plant to follow, which is similar to model reference control.

3.1.2 Design of U-Supertwisting Controller

Supertwisting controller type I:

To integrate supertwisting reference system with U-model control concept, force the nonlinear plant to follow the first equation in the supertwisting reference system (3.1), it gives

$$f(y(k), u(k), \ldots) = \rho_1 x_1(k) + \tau x_2(k) - \tau k_1 |x_1(k)|^{\frac{1}{2}} sign(x_1(k)) \tag{3.5}$$

Supertwisting controller type II:

Similar to U-super twisting controller type I, force the nonlinear plant to follow the second equation of the supertwisting reference system (3.1), it gives

$$f(y(k), u(k), \ldots) = \rho_2 x_2(k) - \tau k_2 sign(x_1(k)) \tag{3.6}$$

Remark 3: It has been proved that the designed control system has the same convergence properties as the constructed supertwisting reference system (3.1).

3.1.3 Simulation Example

Two examples were selected to validate the performance of the proposed procedure and also to illustrate the applications. The numerical experiments were conducted with Matlab simulations.

Example 3.1. Conditional stable nonlinear process model extracted from a heavy oil-cracking furnace as follows:

$$y(k+1) = 0.95y(k) - 0.00408y^3(k) + 0.00408u^3(k) + 0.79 \quad (3.7)$$

The control target is to design a proper controller to drive the process from a given initial state back to origin and the control performance is specified with the supertwisting reference model. First of all, it assigns the parameter to guarantee the stability of the reference model.

$$\rho_1 = 0.5,\ \ \rho_2 = 0.5,\ \ k_1 = 0.5,\ \ k_2 = 0.5,\ \ \tau = 0.1 \quad (3.8)$$

Accordingly, it gives

$$\begin{aligned} x_1(k+1) &= 0.5x_1(k) + 0.1x_2(k) - 0.05|x_1(k)|^{\frac{1}{2}} sign(x_1(k)) \\ x_2(k+1) &= 0.5x_2(k) - 0.05 sign(x_1(k)) \end{aligned} \quad (3.9)$$

For using U-model controller type I, force system (3.7) to follow part of the supertwisting reference model in terms of $x_1(k) = y(k)$. Consequently, a U-model is established with desired output below.

$$\begin{aligned} 0.0408u^3(k) + 0.45x_1(k) - 0.0408y^3(k) + 7.9 - x_2(k) \\ + 0.5|x_1(k)|^{\frac{1}{2}} sign(x_1(k)) = 0 \end{aligned} \quad (3.10)$$

Finally, the controller output $u(k)$ can be determined by resolving one of the three roots (normally the minimal real one) each time. As explained in remark 3, the designed control system consistently stable as specified from the supertwisting reference model.

In simulation the studies, initial output was assigned as $y(0) = 0$. The performance plots are shown in Figs. 3.2–3.4.

To test U-supertwisting controller type II, force nonlinear process (3.7) into $y(k) = x_2(k)$. Consequently, it gives

$$\begin{aligned} &0.95y(k) - 0.00408y^3(k) + 0.00408u^3(k) + 0.79 \\ &= \rho_2 x_2(k) - \tau k_2 sign(x_1(k)) \end{aligned} \quad (3.11)$$

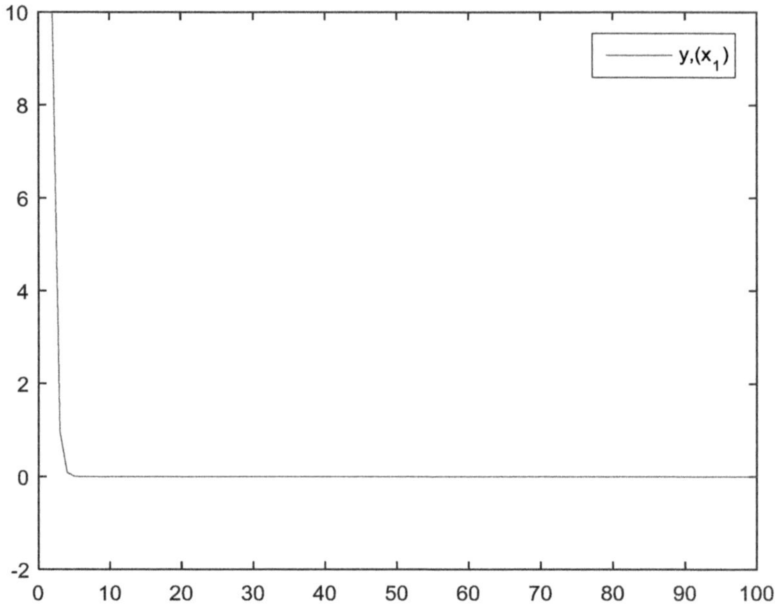

Figure 3.2 Output response of the designed U-model.

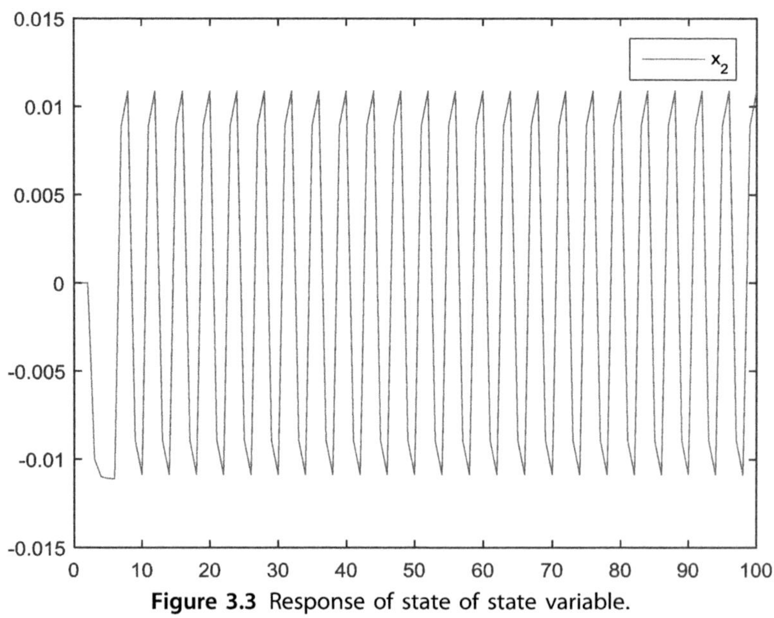

Figure 3.3 Response of state of state variable.

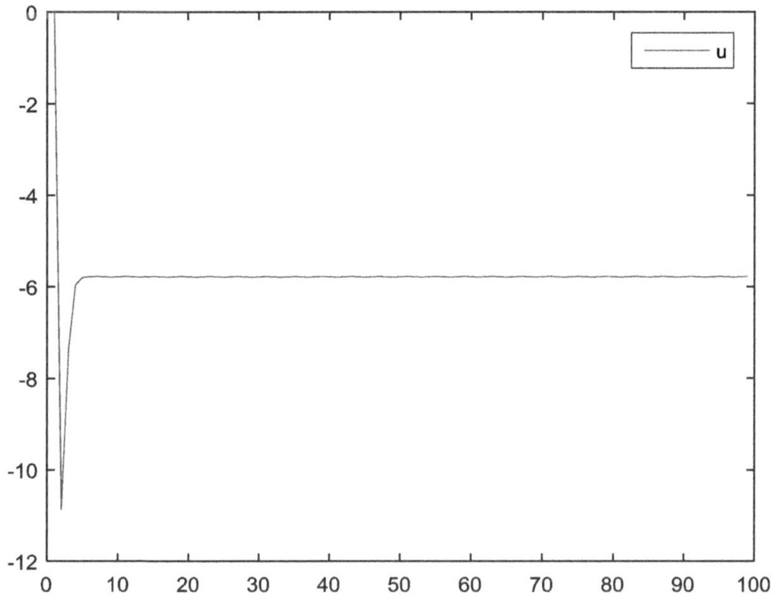

Figure 3.4 Controller output.

The performance plots are shown in Figs. 3.5–3.7.

To test tracking control with controller type I, it assigned $x_1(k) = y(k) - y_d(k)$, where $y_d(k)$ is a reference signal. The performance plots are shown in Figs. 3.8 and 3.9.

Case study 2: Nonlinear rational model

$$y(k+1) = \frac{0.5y(k)u(k) + u^3(k)}{1 + y^2(k) + u^2(k)} \tag{3.12}$$

One of the major advantages using U-model control procedure is to remove the repetition of designing controller while a plant model changed. Thus, the designed supertwisting control system in Case study 1 is still feasible. The sole difference is to form a new U-model from the different plant, which is given by

$$\frac{0.5y(k)u(k) + u^3(k)}{1 + y^2(k) + u^2(k)} = \rho_1 x_1(k) + \tau x_2(k) - \tau k_1 |x_1(k)|^{\frac{1}{2}} sign(x_1(k)) \tag{3.13}$$

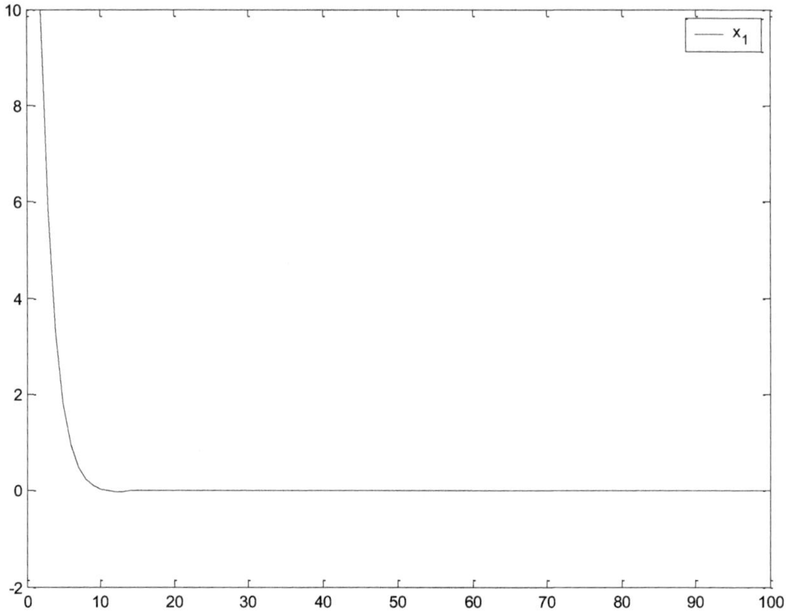

Figure 3.5 Response of state $x_1(k)$.

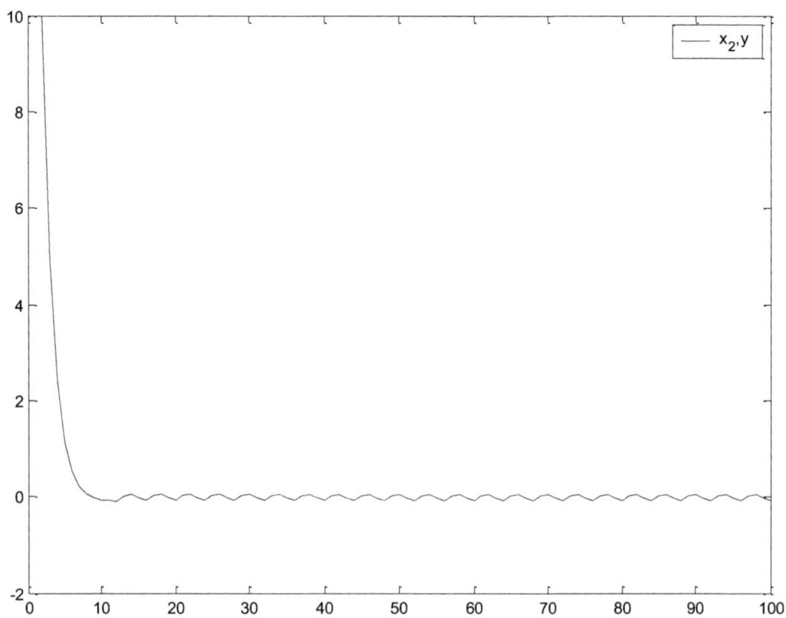

Figure 3.6 Output response.

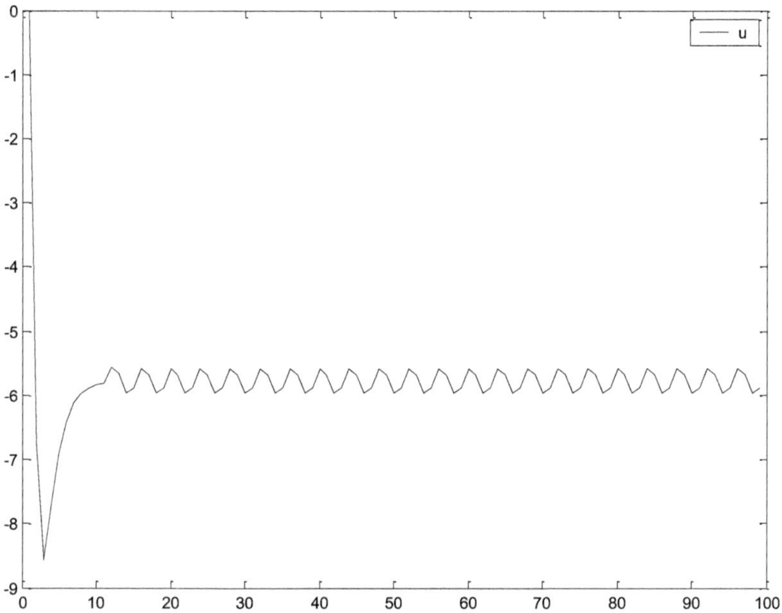

Figure 3.7 Controller output.

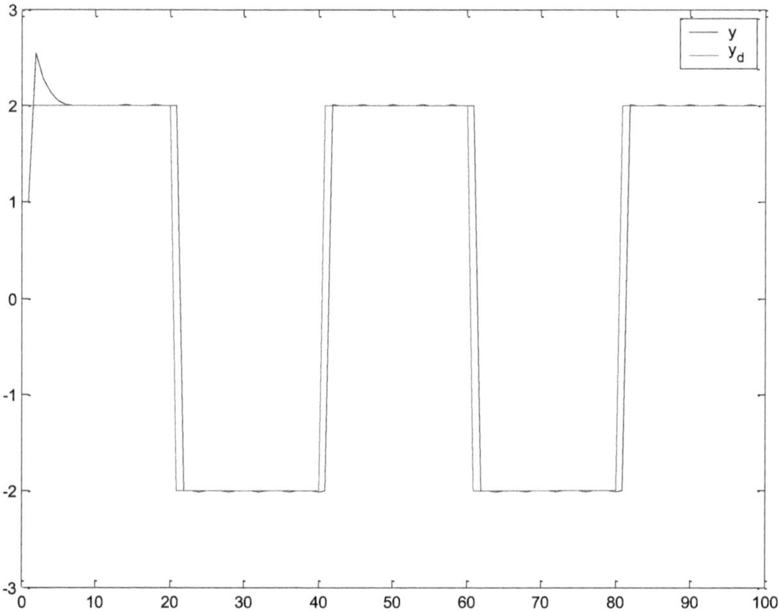

Figure 3.8 Output response to prespecified trajectory.

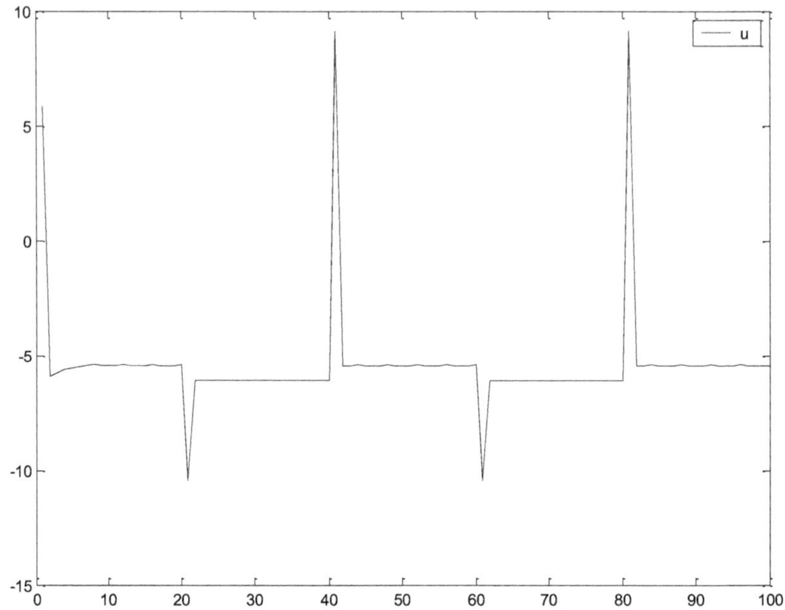

Figure 3.9 Controller output.

The simulated results are shown in Figs. 3.10–3.12.

It should be noted that without U-model approach, it is almost impossible to determine controller output $u(k-1)$ for rational systems, while $u(k-1)$ is appeared in denominator polynomials. Second, it has demonstrated again the generality in once off design of the supertwisting control system and removal of the unnecessary design repetition with the two different plants.

3.2 BACKSTEPPING CONTROL

3.2.1 System Formulation and Preliminaries

Consider the nonlinear strict feedback system (Li et al., 2015; Zhang et al., 2014, 2015).

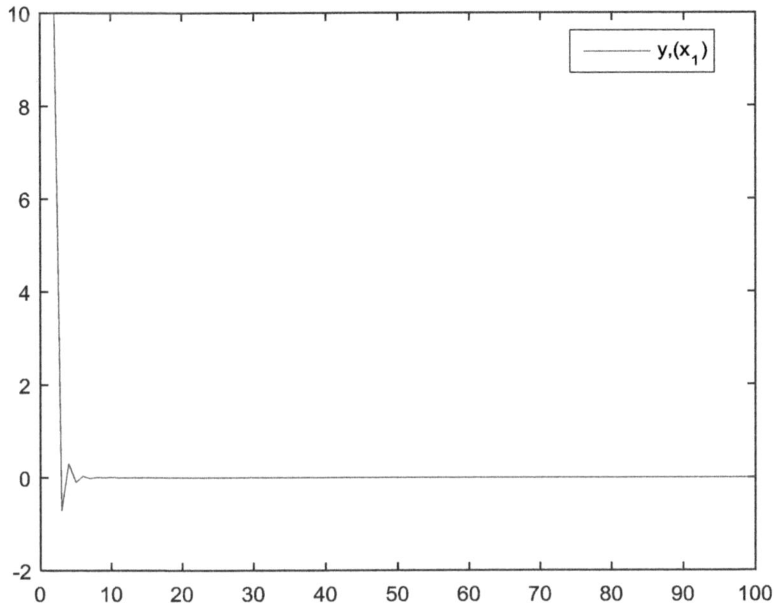

Figure 3.10 Output response of the designed U-model.

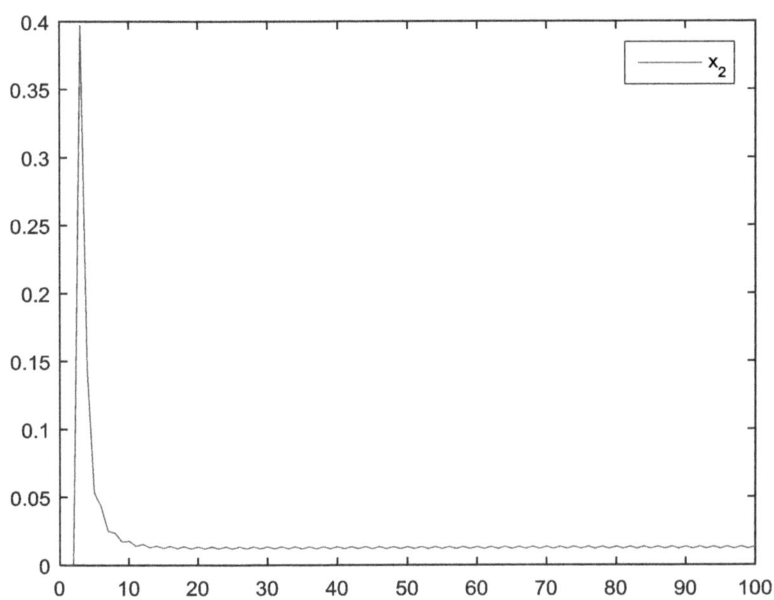

Figure 3.11 Response of state of state variable.

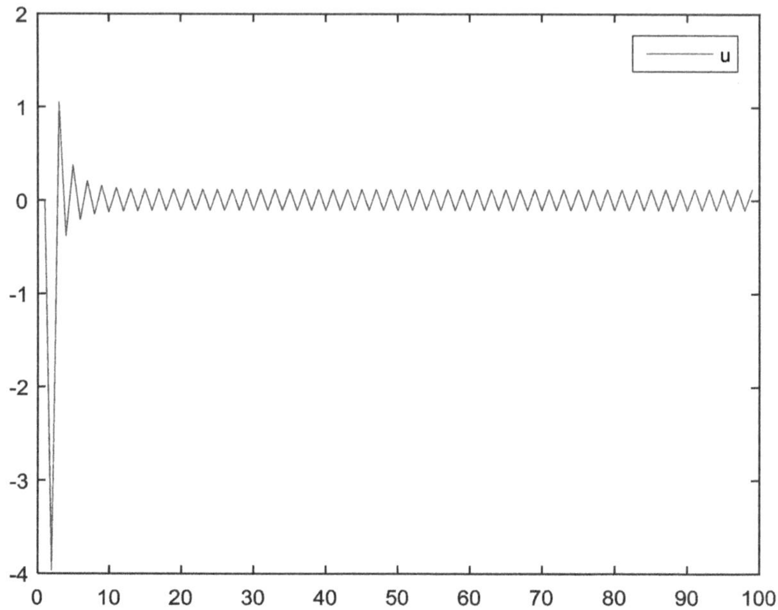

Figure 3.12 Controller output.

$$\begin{cases} \dot{x}_1 = x_2 + f_1(x_1) \\ \dot{x}_2 = x_3 + f_2(x_1, x_2) \\ \vdots \\ \dot{x}_i = x_{i+1} + f_i(x_1, x_2, \ldots, x_i) \\ \vdots \\ \dot{x}_{n-1} = x_n + f_{n-1}(x_1, x_2, \ldots, x_{n-1}) \\ \dot{x}_n = f_n(x_1, x_2, \ldots, x_n) + u \end{cases} \tag{3.14}$$

$$y = x_1$$

where $x = \begin{bmatrix} x_1 & x_2 & \ldots & x_n \end{bmatrix}^T \in R^n$ indicate the state, $u(t) \in R$ and $y(t) \in R$ indicate control input and output, respectively. $f_i(g) : R^i \rightarrow R, i = 1, \ldots, n$ and $f_i(\overline{x}_i) = f_i(x_1, x_2, \ldots, x_i)$ are the control objects, and control system output follows the idea trajectory y_d.

3.2.2 Main Results

The first step, define error state

$$z_1 = x_1 - y_d \tag{3.15}$$

Then,

$$\begin{aligned}\dot{z}_1 &= \dot{x}_1 - \dot{y}_d \\ &= x_2 + f_1(x_1) - \dot{y}_d\end{aligned} \tag{3.16}$$

The idea virtual control law

$$\alpha_1 = -k_1 z_1 - f_1(x_1) + \dot{y}_d \tag{3.17}$$

where $k_1 > 0$,

$$\dot{z}_1 = -k_1 z_1 + x_2 - \alpha_1 \tag{3.18}$$

Define error state

$$z_2 = x_2 - \alpha_1 \tag{3.19}$$

Therefore,

$$\dot{z}_1 = -k_1 z_1 + z_2 \tag{3.20}$$

Choose Lyapunov candidate function

$$V_1 = \frac{1}{2} z_1^2 \tag{3.21}$$

Then,

$$\dot{V}_1 = -k_1 z_1^2 + z_1 z_2 \tag{3.22}$$

The ith $(2 \le i \le n-1)$ step.
Define error state

$$z_i = x_i - \alpha_{i-1} \tag{3.23}$$

Then,

$$\begin{aligned}\dot{z}_i &= \dot{x}_i - \dot{\alpha}_{i-1} \\ &= x_{i+1} + f_i(\overline{x}_i) - \dot{\alpha}_{i-1}\end{aligned} \tag{3.24}$$

The idea virtual control law

$$\alpha_i = -z_{i-1} - k_i z_i - f_i(\overline{x}_i) + \dot{\alpha}_{i-1} \tag{3.25}$$

where $k_i > 0$,

$$\dot{z}_i = -z_{i-1} - k_i z_i + x_{i+1} - \alpha_i \tag{3.26}$$

Define error state

$$z_{i+1} = x_{i+1} - \alpha_i \tag{3.27}$$

Therefore,

$$\dot{z}_i = -z_{i-1} - k_i z_i + z_{i+1} \tag{3.28}$$

Choose Lyapunov candidate function

$$V_i = \frac{1}{2} z_i^2 \tag{3.29}$$

Then,

$$\dot{V}_i = -z_{i-1} z_i - k_i z_i^2 + z_i z_{i+1} \tag{3.30}$$

The nth step, this is the most important step, define error state

$$z_n = x_n - \alpha_{n-1} \tag{3.31}$$

Then,

$$\begin{aligned} \dot{z}_n &= \dot{x}_n - \dot{\alpha}_{n-1} \\ &= f_n(\overline{x}_n) + u - \dot{\alpha}_{n-1} \end{aligned} \tag{3.32}$$

The idea virtual control law

$$u = -z_{n-1} - k_n z_n - f_n(\overline{x}_n) + \dot{\alpha}_{n-1} \tag{3.33}$$

where $k_n > 0$, therefore,

$$\dot{z}_n = -z_{n-1} - k_n z_n \tag{3.34}$$

Choose Lyapunov candidate function

$$V_n = \frac{1}{2} z_n^2 \tag{3.35}$$

Then,

$$\dot{V}_n = -z_{n-1} z_n - k_n z_n^2 \tag{3.36}$$

Choose Lyapunov candidate functions

$$V = V_1 + V_2 + \cdots + V_n \tag{3.37}$$

Then,

$$\dot{V} = -k_1 z_1^2 - k_2 z_2^2 - \cdots - k_n z_n^2 \tag{3.38}$$

Consider the nonlinear pure feedback system

$$\begin{cases} \dot{x}_1 = f_1(x_1, x_2) \\ \dot{x}_2 = f_2(x_1, x_2, x_3) \\ \vdots \\ \dot{x}_i = f_i(x_1, x_2, \ldots, x_{i+1}) \\ \vdots \\ \dot{x}_{n-1} = f_{n-1}(x_1, x_2, \ldots, x_n) \\ \dot{x}_n = f_n(x_1, x_2, \ldots, x_n) + u \end{cases} \tag{3.39}$$

$$y = x_1$$

where $x = [x_1 \quad x_2 \quad \ldots \quad x_n]^T \in R^n$ indicates the state, $u(t) \in R$ and $y(t) \in R$ indicate control input and output, respectively. $f_i(\bullet) : R^i \rightarrow R, i = 1, \ldots, n$ and $f_i(\overline{x}_{i+1}) = f_i(x_1, x_2, \ldots, x_{i+1})$ $f_n(\overline{x}_n) = f_n(x_1, x_2, \ldots, x_n)$ are the control objects and control system output follows the idea trajectory y_d.

The first step, define error state

$$z_1 = x_1 - y_d \tag{3.40}$$

Then,

$$\begin{aligned} \dot{z}_1 &= \dot{x}_1 - \dot{y}_d \\ &= f_1(\overline{x}_2) - \dot{y}_d \end{aligned} \tag{3.41}$$

The idea virtual control law

$$\alpha_1 = -k_1 z_1 + x_2 - f_1(\overline{x}_2) + \dot{y}_d \tag{3.42}$$

where $k_1 > 0$, therefore,

$$\dot{z}_1 = -k_1 z_1 + x_2 - \alpha_1 \tag{3.43}$$

Define error state

$$z_2 = x_2 - \alpha_1 \tag{3.44}$$

Therefore,

$$\dot{z}_1 = -k_1 z_1 + z_2 \tag{3.45}$$

Choose Lyapunov candidate function

$$V_1 = \frac{1}{2} z_1^2 \tag{3.46}$$

Then,

$$\dot{V}_1 = -k_1 z_1^2 + z_1 z_2 \tag{3.47}$$

The ith $(2 \leq i \leq n-1)$ step.

Define error state

$$z_i = x_i - \alpha_{i-1} \tag{3.48}$$

Then,

$$\begin{aligned} \dot{z}_i &= \dot{x}_i - \dot{\alpha}_{i-1} \\ &= f_i(\overline{x}_{i+1}) - \dot{\alpha}_{i-1} \end{aligned} \tag{3.49}$$

The idea virtual control law

$$\alpha_i = -z_{i-1} - k_i z_i + x_{i+1} - f_i(\overline{x}_{i+1}) + \dot{\alpha}_{i-1} \tag{3.50}$$

where $k_i > 0$, therefore,

$$\dot{z}_i = -z_{i-1} - k_i z_i + x_{i+1} - \alpha_i \tag{3.51}$$

Define error state

$$z_{i+1} = x_{i+1} - \alpha_i \tag{3.52}$$

Therefore,

$$\dot{z}_i = -z_{i-1} - k_i z_i + z_{i+1} \tag{3.53}$$

Choose Lyapunov candidate function

$$V_i = \frac{1}{2} z_i^2 \tag{3.54}$$

Then,

$$\dot{V}_i = -z_{i-1} z_i - k_i z_i^2 + z_i z_{i+1} \tag{3.55}$$

The nth step, this is the most important step, define error state

$$z_n = x_n - \alpha_{n-1} \tag{3.56}$$

Then,

$$\begin{aligned} \dot{z}_n &= \dot{x}_n - \dot{\alpha}_{n-1} \\ &= f_n(\overline{x}_n) + u - \dot{\alpha}_{n-1} \end{aligned} \tag{3.57}$$

The control law

$$u = -z_{n-1} - k_n z_n - f_n(\overline{x}_n) + \dot{\alpha}_{n-1} \tag{3.58}$$

where $k_n > 0$, therefore,

$$\dot{z}_n = -z_{n-1} - k_n z_n \tag{3.59}$$

Choose Lyapunov candidate function

$$V_n = \frac{1}{2} z_n^2 \tag{3.60}$$

Then,

$$\dot{V}_n = -z_{n-1} z_n - k_n z_n^2 \tag{3.61}$$

Choose Lyapunov candidate functions

$$V = V_1 + V_2 + \cdots + V_n \tag{3.62}$$

Then,

$$\dot{V} = -k_1 z_1^2 - k_2 z_2^2 - \cdots - k_n z_n^2 \tag{3.63}$$

3.2.3 Simulation Example

In this section, a mathematical example is presented to illustrate the validity of the proposed backstepping control.

$$\begin{aligned} \dot{x}_1 &= 10x_1 - 4x_2 \\ \dot{x}_2 &= x_1 + x_2 + 2x_3 + \sin(x_3) \\ \dot{x}_3 &= x_1^3 + \sin(x_2) + \cos(x_3) + x_2 x_3 + u \\ y &= x_1 \end{aligned}$$

where $x = \begin{bmatrix} x_1 & x_2 & x_3 \end{bmatrix}^T$, u and y are state vector, control input and output, respectively; the reference output is taken as $y_d = \sin(t)$.

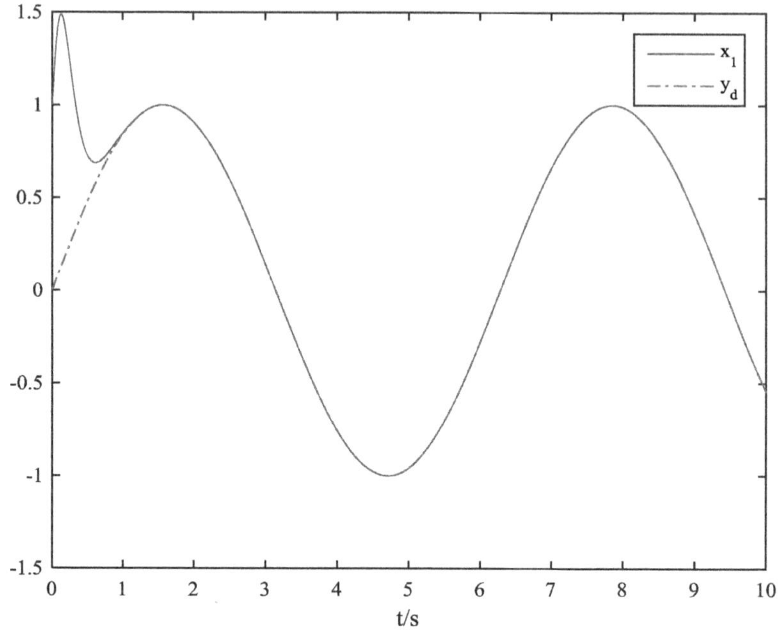

Figure 3.13 Trajectories of x_1, y_d.

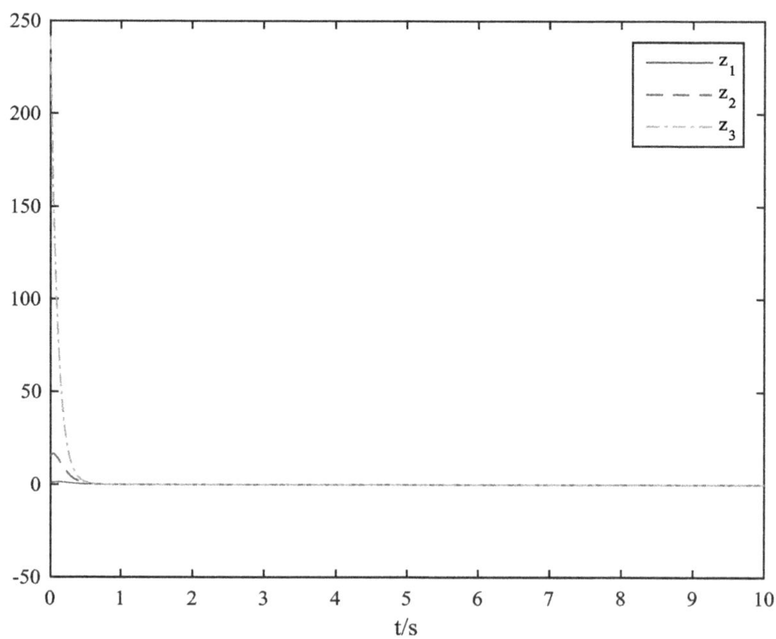

Figure 3.14 Trajectory of error system states.

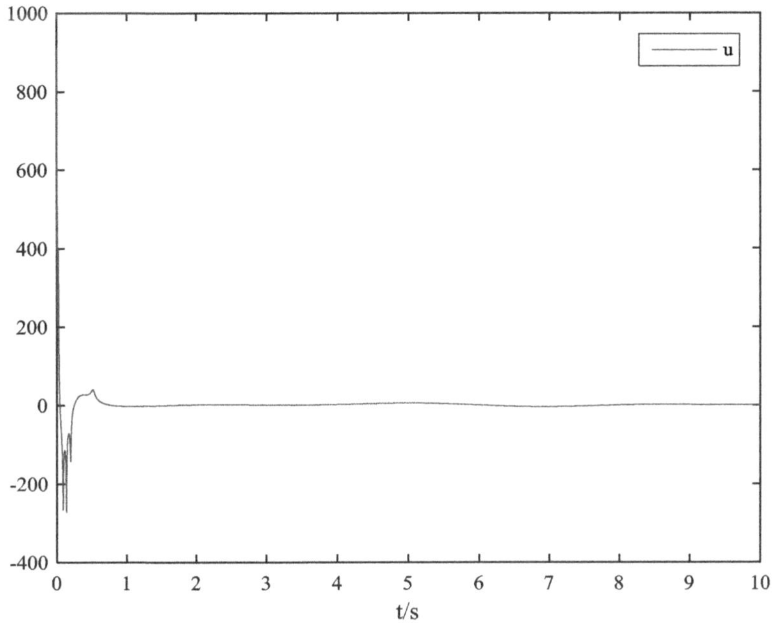

Figure 3.15 Trajectory of control input *u*.

The control objective is to design the tracking control for system such that all the signals in the closed-loop system remain bounded and the system output follows the given reference signal y_d.

In the simulation study, the initial condition choose as $x(0) = [1 \quad 1 \quad 1]^T$, controller gains $k_1 = k_2 = k_3 = 10$. The simulation results are shown in Figs. 3.13–3.15, the output of the designed system satisfies the required accuracy.

REFERENCES

Cheng, C.C., Lin, M.H., Hsiao, J.M., 2000. Sliding mode controllers design for linear discrete-time systems with matching perturbations. Automatica 36, 1205–1211.

Itkis, U., 1976. Control System of Variable Structure.

Lee, D.J., Park, Y., Park, Y.S., 2012. Robust sliding mode descriptor observer for fault and output disturbance estimation of uncertain systems. Automatic Control 57, 2928–2934. IEEE Transactions on.

Li, Y., Wu, X., Zhu, Q., Zhang, J., 2014. Adaptive terminal sliding mode control for finite-time chaos synchronisation with uncertainties and unknown parameters. International Journal of Computer Applications in Technology 49, 340–351.

Li, Y., Zhang, J., Wu, X., 2015. A class of non-affine nonlinear systems adaptive neural networks backstepping control with pure-feedback prototype. Control Conference 1157–1162.

Marino, R., Tomei, P., 1993. Global adaptive output-feedback control of nonlinear systems. II. nonlinear parameterization. IEEE Transactions on Automatic Control 38, 17–32.

Moreno, J.A., Osorio, M., 2012. Strict Lyapunov functions for the super-twisting algorithm. IEEE Transactions on Automatic Control 57, 1035–1040.

Sarpturk, S., Istefanopulos, Y., Kaynak, O., 1987. On the stability of discrete-time sliding mode control systems. IEEE Transactions on Automatic Control 32, 930–932.

Utkin, V., 1977. Variable structure systems with sliding modes. IEEE Transactions on Automatic Control 22, 212–222.

Wu, X.L., 2012. Synchronisation of unified chaotic systems with uncertain parameters in finite time. International Journal of Modelling, Identification and Control 17, 295–301.

Wu, X., Li, Y., Zhang, J., Zhu, Q., 2014. Sliding mode control for neutral systems with uncertain parameters. International Journal of Modelling, Identification and Control 21, 65–71.

Xiaoyuan Luo, S.G., Jiange, W., Xinping, G., 2018. Time delay estimation-based adaptive sliding-mode control for nonholonomic mobile robots. International Journal of Applied Mathematics in Control Engineering 1, 1–8.

ŻAK, S.H., 2003. Systems and Control. Oxford University Press, New York, Oxford.

Zhang, J., Li, Y., Wu, X., 2015. Neural network identification and control for nonlinear dynamic systems with time delay. Control Conference 3524–3529.

Zhang, J., Li, Y., Wu, X., Wang, K., 2014. Adaptive neural network control for pure-feedback nonlinear time delay systems with triangular control structure. Control Conference 1834–1838.

Zhang, J., Zhu, Q., Li, Y., Wu, X., Zhen, R., 2017. A general U-model based super twisting design procedure for nonlinear polynomial systems. Chinese Control Conference 3641–3645.

CHAPTER 4

Advanced Nonlinear Systems Controller Design

Contents

Chapter 4, first, studies chaotic nonlinear systems, and the supertwisting synchronization controller is designed based on U-model method and Lyapunov stability theory, to achieve synchronization between the master and slave systems. Second, it studies the Supertwisting control algorithm used to analyze the nonlinear control system based on U-model. The nonaffine nonlinear systems are described, and the first-order and n-order nonlinear system are designed. Third, it studies discrete nonlinear systems with matching perturbations, the sliding function is proposed, and the nonlinear sliding mode controller is designed

Adaptive Sliding Mode Neural Network Control for Nonlinear Systems
ISBN 978-0-12-815372-7
https://doi.org/10.1016/B978-0-12-815372-7.00004-5

based on Lyapunov stability theory, the controller parameters obtained by matrix technique, the controller input obtained by Newton–Raphson algorithm, to achieve uniformly asymptotic stability of the closed-loop system meanwhile eliminating the chattering effect.

4.1 SUPERTWISTING SYNCHRONIZATION CONTROL OF CHAOTIC SYSTEM–BASED U-MODEL METHOD

4.1.1 Problem Description

In this section, the supertwisting synchronization is analyzed, and the convergence of supertwisting algorithm is proved using Lyapunov function method. The method achieves synchronization between the master and slave systems, and the simulation results show the correctness of the proposed method.

Considered a kind of chaotic systems and pointed out that these systems can be described in a unified form as follows:

$$\begin{cases} \dot{x}_1 = (25\alpha + 10)(x_2 - x_1) \\ \dot{x}_2 = (28 - 35\alpha)x_1 - x_1x_3 + (29\alpha - 1)x_2 \\ \dot{x}_3 = x_1x_2 - \left(\dfrac{8+\alpha}{3}\right)x_3 \end{cases} \tag{4.1}$$

where x_1, x_2 and are state variables and $\alpha \in [0, 1]$. When $0 \leq \alpha < 0.8$, system (Eq. 4.1) is called as the generalized Lorenz chaotic system. When $\alpha = 0.8$, system (Eq. 4.1) is called Lü chaotic system. When $0.8 < \alpha \leq 1$, system (Eq. 4.1) is called generalized Chen chaotic system.

System (Eq. 4.1) is considered as the master system, and the slave system is a controlled system as follows: It is assumed that each state of the master system and the slave system hold uncertainties. Consider the following unified chaotic system with uncertain parameters:

$$\begin{cases} \dot{x}_1 = (25\alpha + 10)(x_2 - x_1) + \Delta_1 \\ \dot{x}_2 = (28 - 35\alpha)x_1 - x_1x_3 + (29\alpha - 1)x_2 + \Delta_2 \\ \dot{x}_3 = x_1x_2 - \left(\dfrac{8+\alpha}{3}\right)x_3 + \Delta_3 \end{cases} \tag{4.2}$$

where Δ_1, Δ_2, and Δ_3 are uncertain parameters.

The slave system is described by

$$\begin{cases} \dot{y}_1 = (25\alpha + 10)(y_2 - y_1) + u_1 \\ \dot{y}_2 = (28 - 35\alpha)y_1 - y_1y_3 + (29\alpha - 1)y_2 + u_2 \\ \dot{y}_3 = y_1y_2 - \left(\dfrac{8+\alpha}{3}\right)y_3 + u_3 \end{cases} \tag{4.3}$$

where y_1, y_2, y_3 are state variables, and u_i is the system control input.

4.1.2 Synchronization Control Based on Supertwisting Algorithm

Supertwisting algorithm is a kind of second-order sliding mode control method; the model can be described as follow:

$$\begin{aligned} \dot{x}_1(t) &= -k_1|x_1|^{\frac{1}{2}} sign(x_1) + x_2 + \rho_1(x,t) \\ \dot{x}_2(t) &= -k_2 sign(x_1) + \rho_2(x,t) \end{aligned} \tag{4.4}$$

where x_1, x_2 are the state variables, k_1, k_2 are the control gain, and ρ_1, ρ_2 are the bounded perturbations. Consider the system without perturbations, $\rho_1 = \rho_2 = 0$ choose $\varsigma^T = [x_1 \quad x_2]$ $= \left[|x_1|^{\frac{1}{2}} sign(x_1) \quad x_2 \right]$, then

$$\dot{\varsigma} = \frac{1}{|\varsigma_1|}A\varsigma, A = \begin{bmatrix} -\frac{1}{2}k_1 & \frac{1}{2} \\ -k_2 & 0 \end{bmatrix} \tag{4.5}$$

Choose Lyapunov function as follow:

$$V = \varsigma^T P \varsigma \tag{4.6}$$

Take derivative of the Lyapunov function as follow:

$$\dot{V} = -|x_1|^{-\frac{1}{2}} \varsigma^T Q \varsigma \tag{4.7}$$

The matrix Q satisfies the equation as follow:

$$A^T P + PA = -Q \tag{4.8}$$

Consider the system without perturbations, $\rho_1 = \rho_2 = 0$, then we have the equivalence of four condition:

The system can reach the equilibrium point $x = 0$ in finite time;

The matrix A is Hurwitz matrix, that is, all the real parts of the eigen values are in the left half plane;

The control gains are positive, $k_1 > 0, k_2 > 0$;

For any positive definite symmetric matrix $Q = Q^T > 0$, there is a positive definite symmetric solution $P = P^T > 0$ for the Lyapunov Eq. (4.8). For function $V(x)$, the trajectories of the solution will reach the original point $x(0) = x_0$ within a limited time $T(x_0)$, and the following relation is established:

$$T(x_0) = \frac{2}{\sigma} V^{\frac{1}{2}}(x_0), \sigma = \frac{\lambda_{\min}^{\frac{1}{2}}(P)\lambda_{\min}(Q)}{\lambda_{\max}(P)}$$

The study of convergence shows that the system is finite time convergent and has good robustness.

Theorem 4.1: Consider the master chaotic system (Eq. 4.1) and the slave chaotic system (Eq. 4.3); for any constant $k_i > 0, l_i > 0, (i = 1,2,3)$, we have the following supertwisting controller:

$$\begin{cases} u_1 = -(25\alpha + 10)(e_2 - e_1) - k_1|e_1|^{\frac{1}{2}} sign(e_1) + v_1 \\ u_2 = -(28 - 35\alpha)e_1 - (29\alpha - 1)e_2 \\ -e_1 e_3 + e_1 y_3 + e_3 y_1 - k_2|e_2|^{\frac{1}{2}} sign(e_2) + v_2 \\ u_3 = \left(\frac{8+\alpha}{3}\right) e_3 + e_1 e_2 - e_1 y_2 - e_2 y_1 - k_3|e_3|^{\frac{1}{2}} sign(e_3) + v_3 \end{cases}$$

where

$$\begin{cases} \dot{v}_1 = -l_1 sign(e_1) \\ \dot{v}_2 = -l_2 sign(e_2) \\ \dot{v}_3 = -l_3 sign(e_3) \end{cases} \tag{4.10}$$

Then the master chaotic system (Eq. 4.1) and slave chaotic system (Eq. 4.3) can synchronize in finite time.

Proof: The error of the master and slave system can be described as follow:

$$e_i = y_i - x_i, (i = 1, 2, 3) \tag{4.11}$$

That is,

$$\begin{cases} \dot{e}_1 = (25\alpha + 10)(e_2 - e_1) + u_1 \\ \dot{e}_2 = (28 - 35\alpha)e_1 + (29\alpha - 1)e_2 \\ \quad +e_1e_3 - e_1y_3 - e_3y_1 + u_2 \\ \dot{e}_3 = -\left(\dfrac{8+\alpha}{3}\right)e_3 - e_1e_2 + e_1y_2 + e_2y_1 + u_3 \end{cases} \tag{4.12}$$

Substitute the U-based controller Eq. (4.9) into Eq. (4.12)

$$\begin{cases} \dot{e}_1 = -k_1|e_1|^{\frac{1}{2}}sign(e_1) + v_1 \\ \dot{e}_2 = -k_2|e_2|^{\frac{1}{2}}sign(e_2) + v_2 \\ \dot{e}_3 = -k_3|e_3|^{\frac{1}{2}}sign(e_3) + v_3 \end{cases} \tag{4.13}$$

where

$$\begin{cases} \dot{v}_1 = -l_1 sign(e_1) \\ \dot{v}_2 = -l_2 sign(e_2) \\ \dot{v}_3 = -l_3 sign(e_3) \end{cases} \tag{4.14}$$

The error system (Eq. 4.13) can be convergence in finite according to supertwisting algorithm.

The synchronization control algorithm has strong robustness. If we want to synchronize the chaotic system (Eq. 4.2) with uncertain external disturbances, we also use the synchronization controller in Theorem 4.1, and the error system is expressed as:

$$\begin{cases} \dot{e}_1 = -k_1|e_1|^{\frac{1}{2}}sign(e_1) + v_1 + \Delta_1 \\ \dot{e}_2 = -k_2|e_2|^{\frac{1}{2}}sign(e_2) + v_2 + \Delta_2 \\ \dot{e}_3 = -k_3|e_3|^{\frac{1}{2}}sign(e_3) + v_3 + \Delta_3 \end{cases} \tag{4.15}$$

where

$$\begin{cases} \dot{v}_1 = -l_1 sign(e_1) \\ \dot{v}_2 = -l_2 sign(e_2) \\ \dot{v}_3 = -l_3 sign(e_3) \end{cases} \tag{4.16}$$

The system is supertwisting finite time stable if the perturbations Δ_i,$i = 1,2,3$ are unknown constants. If the disturbance is time varying, as long as meet $|\dot{\Delta}_i| < \rho, i = 1, 2, 3$, ρ is known constant, then according to the finite time stability characteristics of super twisting system, it can be proved that the system is finite time convergence.

4.1.3 Simulation Example

Example 4.1 Consider the master chaotic system as follow:

$$\begin{cases} \dot{x}_1 = (25\alpha + 10)(x_2 - x_1) \\ \dot{x}_2 = (28 - 35\alpha)x_1 - x_1x_3 + (29\alpha - 1)x_2 \\ \dot{x}_3 = x_1x_2 - \left(\dfrac{8+\alpha}{3}\right)x_3 \end{cases} \tag{4.17}$$

Consider the slave chaotic system as follow:

$$\begin{cases} \dot{y}_1 = (25\alpha + 10)(y_2 - y_1) + u_1 \\ \dot{y}_2 = (28 - 35\alpha)y_1 - y_1y_3 + (29\alpha - 1)y_2 + u_2 \\ \dot{y}_3 = y_1y_2 - \left(\dfrac{8+\alpha}{3}\right)y_3 + u_3 \end{cases} \tag{4.18}$$

Choose the controller in Theorem 4.1, and then, we have the following results:

Choose the initial condition as

$$\begin{array}{l} (\,x_1(0)\quad x_2(0)\quad x_3(0)\,)^T = (\,-5\quad 1\quad 5\,)^T \\ (\,y_1(0)\quad y_2(0)\quad y_3(0)\,)^T = (\,0\quad 0\quad 0\,)^T \\ (\,v_1(0)\quad v_2(0)\quad v_3(0)\,)^T = (\,1\quad 1\quad 1\,)^T \end{array} \tag{4.19}$$

Choose the control gain as $k_1 = 1, k_2 = 1, k_3 = 5$,

$$\begin{cases} \dot{v}_1 = -sign(e_1) \\ \dot{v}_2 = -sign(e_2) \\ \dot{v}_3 = -5sign(e_3) \end{cases} \tag{4.20}$$

Then, we have,

$$A = \begin{bmatrix} -\frac{1}{2} & \frac{1}{2} \\ -1 & 0 \end{bmatrix}, P = \begin{bmatrix} 1.9236 & -0.3516 \\ -0.3516 & 1.2617 \end{bmatrix}, Q = \begin{bmatrix} 1.2204 & 0.1241 \\ 0.1241 & 0.3516 \end{bmatrix}$$

$$\lambda(P) = (\,1.1098\quad 2.0756\,), \lambda(Q) = (\,0.3343\quad 1.2378\,)$$

So we have $T = 79.8132$, the finite time has relationship with the initial conditions. Figs. 4.1 and 4.2 describes the chaotic characteristic of Lorenz chaotic system; Fig. 4.3 describes the master chaotic system state x_1 and slave chaotic system state y_1; Fig. 4.4 describes the master chaotic system state x_2 and slave chaotic system state y_2; Fig. 4.5 describes the master

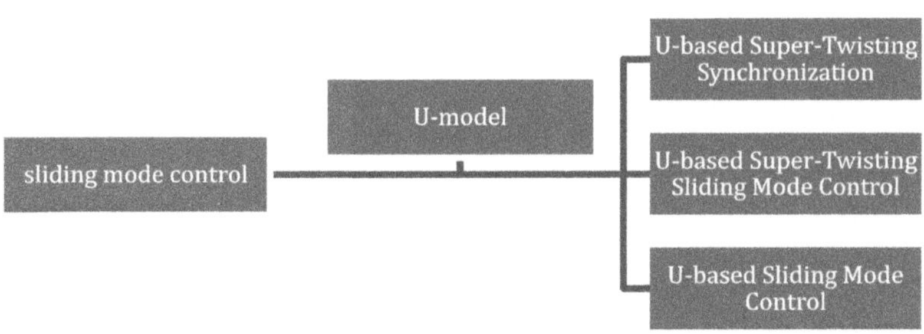

Figure 4.1 The Structure Chart of Chapter 4.

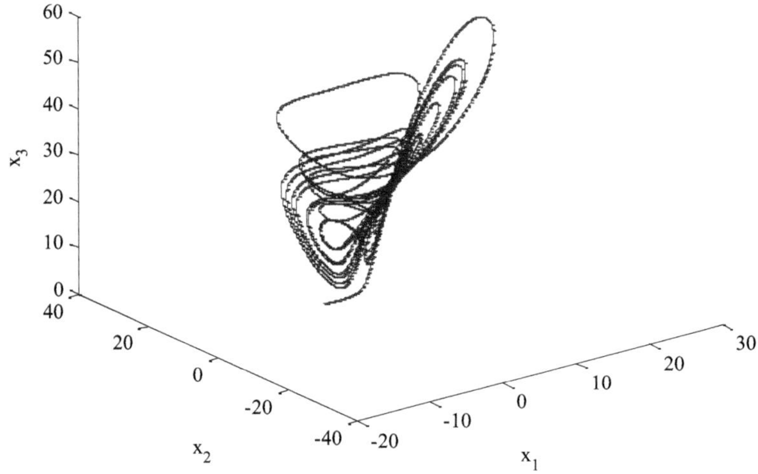

Figure 4.2 Lorenz chaotic system.

chaotic system state x_3 and slave chaotic system state y_3; and Fig. 4.6 shows the state synchronization error. It can be seen from the simulation results that the state of the slave chaotic system is synchronized to the state of the master chaotic system after the controller is added to the system.

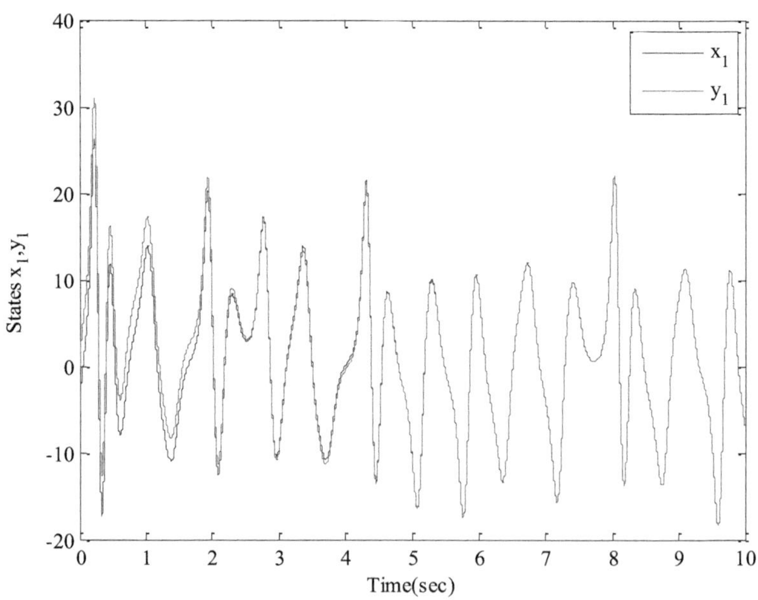

Figure 4.3 Lorenz chaotic system x_1.

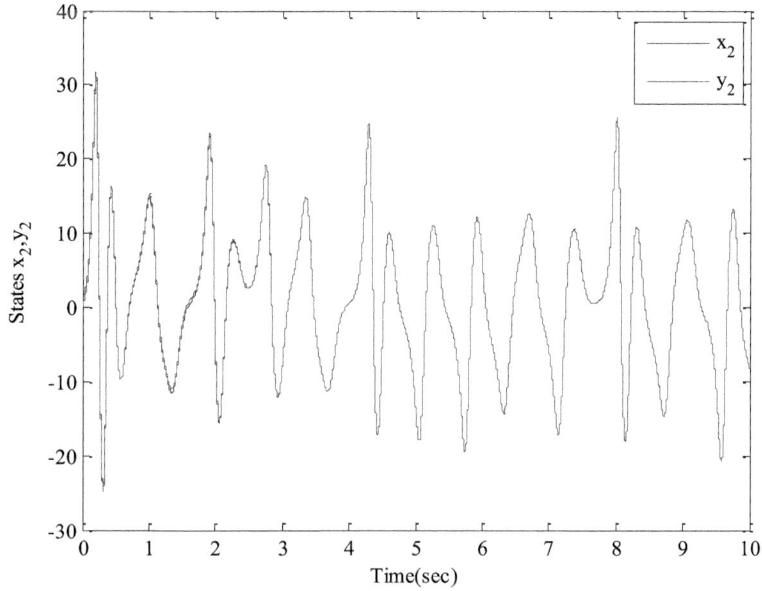

Figure 4.4 Lorenz chaotic system x_2.

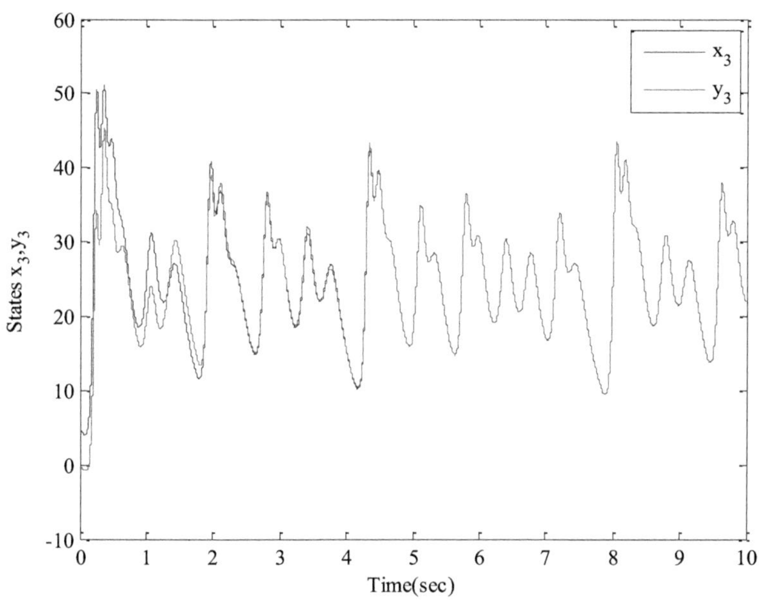

Figure 4.5 Lorenz chaotic system x_3.

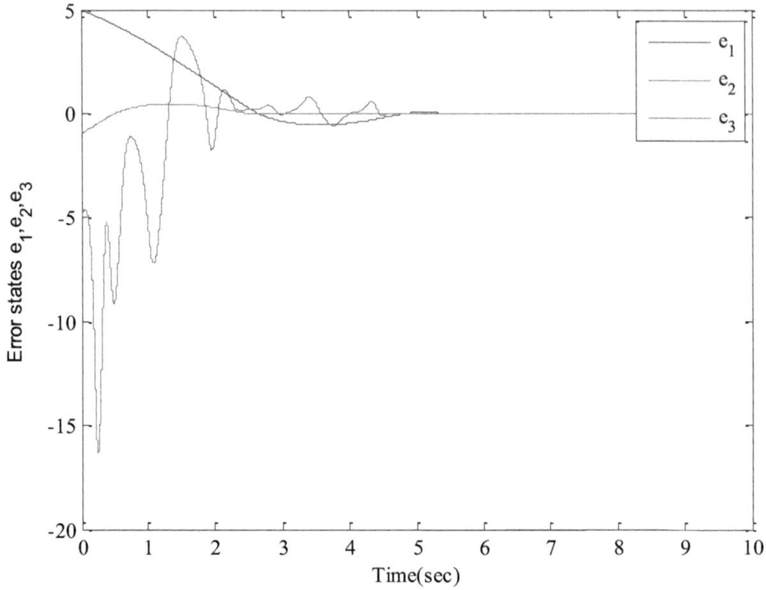

Figure 4.6 The synchronization error.

Example 4.2 Consider the Chen chaotic system, choose the controller in Theorem 4.1, then, we have the following results:

Choose the initial condition as

$$\begin{array}{l} (x_1(0) \quad x_2(0) \quad x_3(0))^T = (-5 \quad 1 \quad 5)^T \\ (y_1(0) \quad y_2(0) \quad y_3(0))^T = (0 \quad 0 \quad 0)^T \\ (v_1(0) \quad v_2(0) \quad v_3(0))^T = (1 \quad 1 \quad 1)^T \end{array} \tag{4.21}$$

Choose the control gain as $k_1 = 1, k_2 = 1, k_3 = 5$,

$$\begin{cases} \dot{v}_1 = -sign(e_1) \\ \dot{v}_2 = -sign(e_2) \\ \dot{v}_3 = -5sign(e_3) \end{cases} \tag{4.22}$$

Fig. 4.7 describes the chaotic characteristic of Chen chaotic system; Fig. 4.8 describes the master chaotic system state x_1 and slave chaotic system state y_1; Fig. 4.9 describes the master chaotic system state x_2 and slave chaotic system state y_2; Fig. 4.10 describes the master chaotic system state x_3 and slave chaotic system

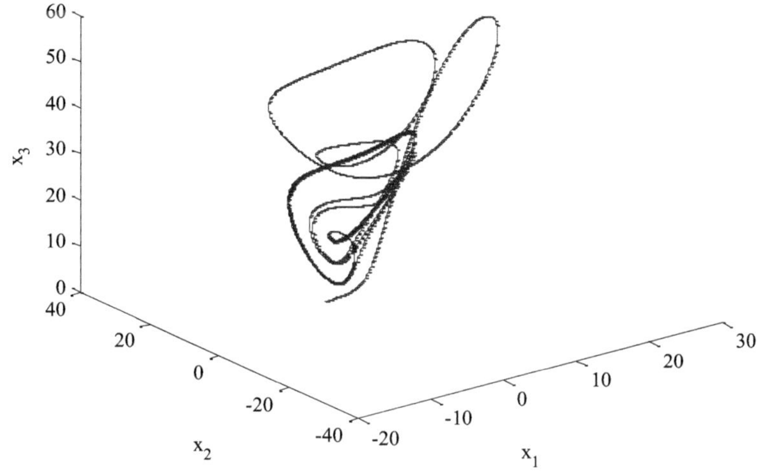

Figure 4.7 Chen chaotic system.

state y_3; and Fig. 4.11 shows the state synchronization error. It can be seen from the simulation results that the state of the slave chaotic system is synchronized to the state of the master chaotic system after the controller is added to the system.

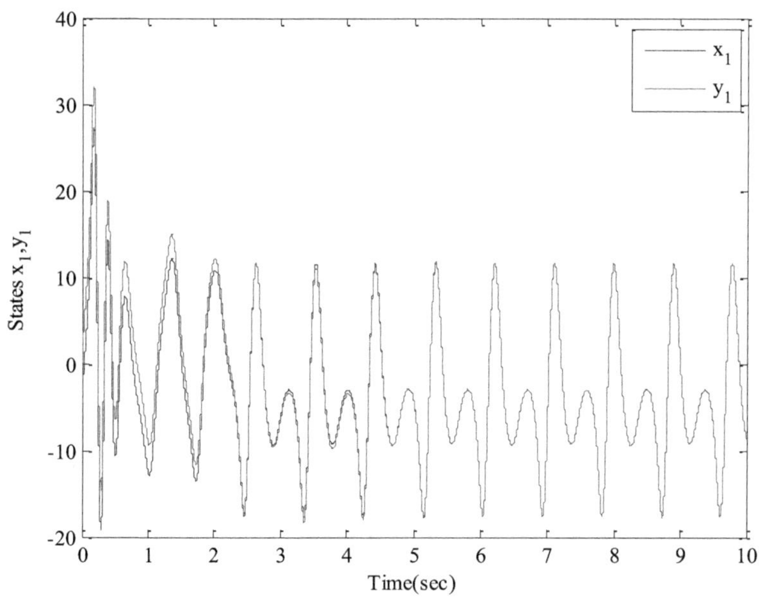

Figure 4.8 Chen chaotic system x_1.

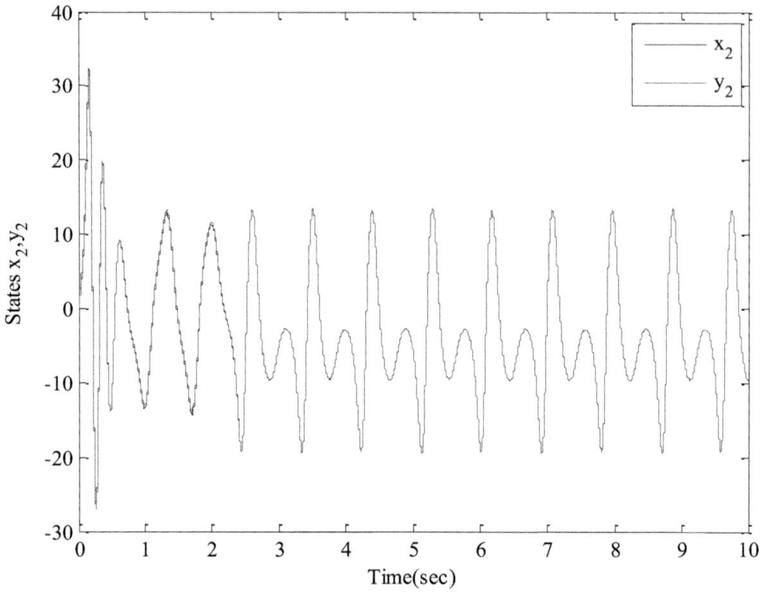

Figure 4.9 Chen chaotic system x_2.

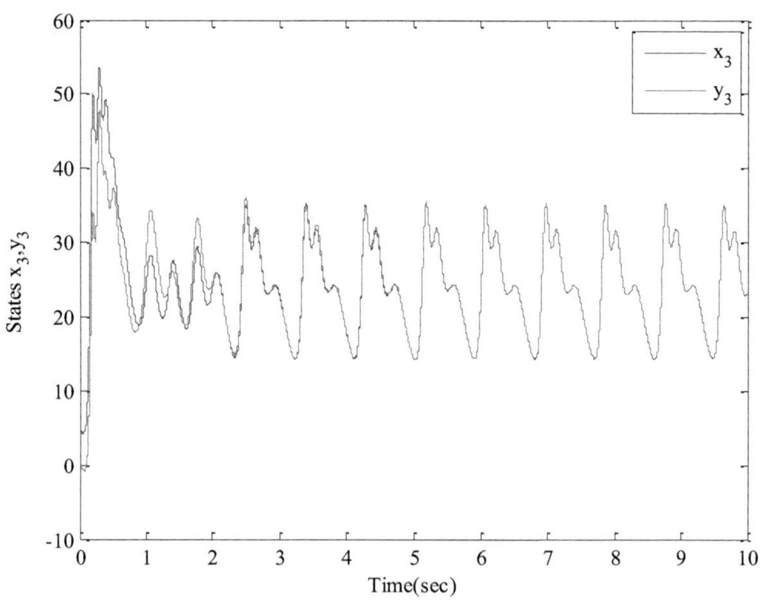

Figure 4.10 Chen chaotic system x_3.

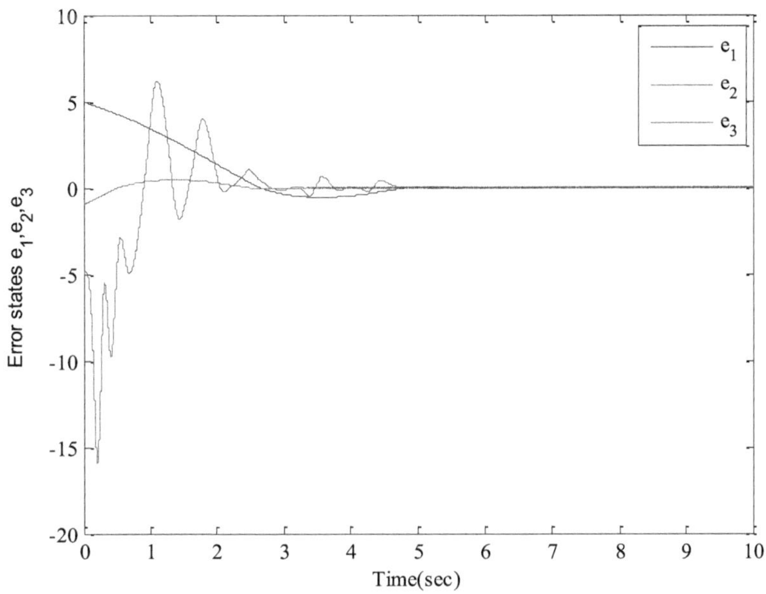

Figure 4.11 The synchronization error.

4.2 SUPERTWISTING SLIDING MODE CONTROL OF NONLINEAR SYSTEM—BASED U-MODEL METHOD

In this section, the nonlinear control system based on U-model is analyzed, and the convergence of the supertwisting algorithm is proved using Lyapunov function method. The output of the system is simulated, and the simulation results show the correctness of the proposed method.

4.2.1 Problem Description

Consider the following nonaffine nonlinear systems:

$$\begin{aligned} \dot{x}_i(t) &= h_i(\overline{x}_i(t)) + x_{i+1}(t), 1 \le i \le n-1 \\ \dot{x}_n(t) &= h_n(\overline{x}_n(t), u(t)) \\ y(t) &= x_1(t) \end{aligned} \tag{4.23}$$

where $x_i \in \Re$, $y \in \Re$, and $u \in \Re$, respectively, are the state of the system, control output, and control input; $y_d \in \Re$ is a predetermined control target; the purpose of the control is to make the system output tracking system control objectives, $\overline{x}_i = [x_1, x_2, \ldots, x_i]^T \in \Re^i$, where $i = 1,2,\ldots n$, and h_i represents a nonlinear smooth function.

$$h(x(t)) = W^* h_N(x(t)) - \varepsilon(t) \tag{4.24}$$

Neural network $W^* h_N(x(t))$ is used to approximate the nonlinear function, $\varepsilon(t)$ is the approximation error, and here W^* is the ideal approximation of nonlinear system neural network.

$$W^* = \arg\min_{W \in \Re^l} \{\sup|h(x) - W h_N(x)|\} \tag{4.25}$$

Here, W^*needs to be estimated with $\widehat{W}$ and through the algorithm to adjust. For the matrix of the estimated value and the ideal value, using $\widetilde{W} = \widehat{W} - W^*$ and the function of neural network, the bounded approximation error of the nonlinear system is $\varepsilon(t)$, and $|\varepsilon(t)|^2 < \varepsilon$。

Consider the system (Eq. 4.23), select the following coordinate transformation

$$\begin{gathered} z_1 = x_1 - y_d \\ z_i = x_i - \alpha_{i-1}, i = 2, 3, \ldots n \end{gathered} \tag{4.26}$$

where y_d represents the desired output of the system, the α_i used as the virtual controller, and the ideal control law is

$$\begin{gathered} \alpha_1^* = -k_1 z_1 - h_1 + \dot{y}_d \\ \alpha_i^* = -z_{i-1} - k_i z_i - h_i + \dot{\alpha}_{i-1}^* \\ 2 \leq i \leq n-1 \\ \alpha_{n-1}^* = -z_{n-2} - k_{n-1} z_{n-1} - h_{n-1} + \dot{\alpha}_{n-2}^* \end{gathered} \tag{4.27}$$

4.2.2 One-Order Nonlinear System Control

One-order nonaffine nonlinear systems are represented as

$$\dot{x}(t) = h(x(t), u(t)) \tag{4.28}$$

Here, the function $h(x)$ represents the nonlinear function.

Theorem 4.1 According to the adaptive algorithm to adjust the weights of the neural network, the closed-loop system with bounded initial conditions is bounded.

Prove: According to the system can be obtained

$$z_1 = x - y_d \tag{4.29}$$

We can get

$$\dot{z} = h(x, u) - \dot{y}_d \tag{4.30}$$

The virtual control rate is selected as the

$$\alpha^* = -k_1|z_1|^{\frac{1}{2}} sign(z_1) - h(x, u) + \dot{y}_d + u \tag{4.31}$$

We can get

$$\dot{z}_1 = -k_1|z_1|^{\frac{1}{2}} sign(z_1) + z_2 \tag{4.32}$$

If the assumptions are as follows

$$z_2 = u - \alpha^* \tag{4.33}$$

we can get

$$\dot{z}_2 = \gamma - \dot{\alpha}^* \tag{4.34}$$

We can select

$$\gamma = \dot{\alpha}^* - k_2 sign(z_1) \tag{4.35}$$

We can get

$$\dot{z}_2 = -k_2 sign(z_1) \tag{4.36}$$

To meet the twisting super control algorithm, the following neural network approximation of nonlinear function is used, so there is

$$h(x, u) = \omega^{*T} \psi(x, u) + \varepsilon \tag{4.37}$$

So select the virtual control law

$$\alpha = -k_1|z_1|^{\frac{1}{2}} sign(z_1) - \widehat{\omega}^T \psi(x, u) + \dot{y}_d + u \tag{4.38}$$

The neural network is expressed as

$$\dot{\widehat{\omega}} = \tau_1^{-1} sign(z_1) \psi(x, u) \tag{4.39}$$

System can be expressed as

$$\dot{z}_1 = \omega^{*T}\psi(x,u) + \varepsilon - \dot{y}_d - k_1|z_1|^{\frac{1}{2}}sign(z_1) - \widehat{\omega}^T\psi(x,u) + \dot{y}_d + u - \alpha \tag{4.40}$$

We can get

$$\dot{z}_1 = -k_1|z_1|^{\frac{1}{2}}sign(z_1) - \widetilde{\omega}\psi(x,u) + z_2 + \varepsilon \tag{4.41}$$

Select neural network

$$\dot{\alpha}^* = \omega_u^{*T}\psi(x,u) \tag{4.42}$$

The selected virtual control rate

$$v = \widehat{\omega}_u^T\psi(x,u) - k_2 sign(z_2) \tag{4.43}$$

We can get

$$\begin{aligned}\dot{z}_2 &= \widehat{\omega}_u^T\psi(x,u) - k_2 sign(z_1) - \omega_u^{*T}\psi(x,u)\\ &= -\widetilde{\omega}_u^T\psi(x,u) - k_2 sign(z_1)\end{aligned} \tag{4.44}$$

The convergence of the system is proved, and the Lyapunov functional is selected as the

$$V_1 = k_2|z_1| + \frac{1}{2}z_2^2 \tag{4.45}$$

We can get

$$\dot{V}_1 = -k_1k_2|z_1|^{\frac{1}{2}} - k_2 sign(z_1)\widetilde{\omega}\psi(x,u) + k_2 sign(z_1)\varepsilon - z_2\widetilde{\omega}_u^T\psi(x,u) \tag{4.46}$$

Neural network weight correction algorithm

$$\dot{\widehat{\omega}} = \tau_1^{-1}sign(z_1)\psi(x,u) \tag{4.47}$$

Lyapunov functional for the part of neural network

$$V_2 = \frac{k_2}{2}\widetilde{\omega}^T\tau_1\widetilde{\omega} + \frac{1}{2}\widetilde{\omega}_u^T\tau_2\widetilde{\omega}_u \tag{4.48}$$

We can get

$$\dot{V}_2 = k_2\widetilde{\omega}^T sign(z_1)\psi(x,u) + \widetilde{\omega}_u^T z_1\psi(x,u) \tag{4.49}$$

For the system, the design is as follows

$$V = V_1 + V_2 \tag{4.50}$$

We can get

$$\dot{V} \le -k_1k_2|z_1|^{\frac{1}{2}} + k_2 sign(|z_1|) \tag{4.51}$$

Theorem is proved.

4.2.3 N-order Nonlinear System Control

N-order nonaffine nonlinear system, the design of virtual controller to achieve the inverse iteration control.

First step:

$$z_1 = x_1 - y_d \tag{4.52}$$

According to the system,

$$\dot{z}_1(t) = h_1(x_1(t)) + x_2(t) - \dot{y}_d \tag{4.53}$$

The ideal virtual control law is as follows

$$\alpha_1^* = -k_1z_1 - h_1 + \dot{y}_d \tag{4.54}$$

The neural network is selected to approximate the nonlinear $h_1(x_1(t))$, as follows

$$h_1(x_1(t)) = W_1^* h_N(x_1(t)) - \varepsilon_1(t) \tag{4.55}$$

The virtual control law is selected as follows

$$\alpha_1 = -k_1z_1 - \widehat{W}_1 h_N(x_1(t)) + \dot{y}_d \tag{4.56}$$

So we can get

$$\dot{z}_1 = -\widetilde{W}_1 h_N(x_1(t)) - k_1z_1 + z_2 - \varepsilon_1(t) \tag{4.57}$$

and

$$\widetilde{W}_1 = \widehat{W}_1 - W_1^* \tag{4.58}$$

For subsystem

$$x_2 = \alpha_1 + z_2 \tag{4.59}$$

The following adaptive laws are selected as follows

$$\dot{\widehat{W}}_1^{\mathrm{T}} = \Gamma_1^{-1}\left[z_1 h_N(x_1) - \sigma_1 \widehat{W}_1^{\mathrm{T}}\right] \tag{4.60}$$

Select Lyapunov function as

$$V_1 = \frac{1}{2}z_1^2 + \frac{1}{2}\widetilde{W}_1 \Gamma_1 \widetilde{W}_1^{\mathrm{T}} \tag{4.61}$$

The system of the time derivative can be as follows

$$\dot{V}_1 \leq -\left(k_1 - \frac{1}{2}\right)z_1^2 + z_1 z_2 + \frac{1}{2}\varepsilon_1 - \frac{\sigma_1}{2}\widetilde{W}_1 \widetilde{W}_1^{\mathrm{T}} + \frac{\sigma_1}{2}W_1^* W_1^{*\mathrm{T}} \tag{4.62}$$

The system of the time derivative and apply the following inequality

$$-\sigma_1 \widetilde{W}_1 \widehat{W}_1^{\mathrm{T}} \leq -\frac{\sigma_1}{2}\widetilde{W}_1 \widetilde{W}_1^{\mathrm{T}} + \frac{\sigma_1}{2}W_1^* W_1^{*\mathrm{T}} \tag{4.63}$$

According to the above process, it can be as follows:

$$\dot{V}_1 \leq -\left(k_1 - \frac{1}{2}\right)z_1^2 + z_1 z_2 - \eta_1 V_1 + \delta_1 \tag{4.64}$$

where

$$\begin{gathered} k_1 > -\frac{1}{2} \\ \eta_1 = \frac{\sigma_1}{\lambda_{\max}(\Gamma_1)} \\ \delta_1 = \frac{\sigma_1}{2}W_1^* W_1^{*\mathrm{T}} + \frac{1}{2}\varepsilon_1 \end{gathered} \tag{4.65}$$

Step i, $(2 \leq i \leq n-1)$

$$z_i = x_i - \alpha_{i-1} \tag{4.66}$$

According to the system

$$\dot{z}_i = h_i(\overline{x}_i(t)) - \dot{\alpha}_{i-1} + x_{i+1} \tag{4.67}$$

Ideal virtual control law

$$\alpha_i^* = -z_{i-1} - k_i z_i - h_i + \dot{\alpha}_{i-1} \tag{4.68}$$

The neural network is selected to approximate the nonlinear $h_i - \dot{\alpha}_{i-1}$

$$h_i(\overline{x}_i(t)) - \dot{\alpha}_{i-1} + \frac{\partial \alpha_i}{\partial y_d}\dot{y}_d = W_i^* h_N(\overline{x}_i(t)) - \varepsilon_i(t) \tag{4.69}$$

Select control law as

$$\alpha_i = -z_{i-1} - k_i z_i - \widehat{W}_i h_N(\overline{x}_i(t)) + \frac{\partial \alpha_i}{\partial y_d}\dot{y}_d \tag{4.70}$$

Then,

$$\dot{z}_i = -\widetilde{W}_i h_N(\overline{x}_i(t)) - z_{i-1} - k_i z_i + z_{i+1} - \varepsilon_i(t) \tag{4.71}$$

where

$$\widetilde{W}_i = \widehat{W}_i - W_i^* \tag{4.72}$$

Select the following adaptive law

$$\dot{\widehat{W}}_i^{\mathrm{T}} = \Gamma_i^{-1}\left[z_i h_N(\overline{x}_i) - \sigma_i \widehat{W}_i^{\mathrm{T}}\right] \tag{4.73}$$

Select Lyapunov function

$$V_i = \frac{1}{2}z_i^2 + \frac{1}{2}\widetilde{W}_i \Gamma_i \widetilde{W}_i^{\mathrm{T}} \tag{4.74}$$

The system of the time derivative can be

$$\dot{V}_i \leq -z_{i-1}z_i - \left(k_i - \frac{1}{2}\right)z_i^2 + z_i z_{i+1} + \frac{1}{2}\varepsilon_i - \frac{\sigma_i}{2}\widetilde{W}_i \widetilde{W}_i^{\mathrm{T}} + \frac{\sigma_i}{2}W_i^* W_i^{*\mathrm{T}} \tag{4.75}$$

The system of the time derivative and apply the following inequality

$$-\sigma_i \widetilde{W}_i \widehat{W}_i^{\mathrm{T}} \leq -\frac{\sigma_i}{2} \widetilde{W}_i \widetilde{W}_i^{\mathrm{T}} + \frac{\sigma_i}{2} W_i^* W_i^{*\mathrm{T}} \tag{4.76}$$

We can get

$$\dot{V}_i \leq -z_{i-1} z_i - \left(k_i - \frac{1}{2}\right) z_i^2 + z_i z_{i+1} - \eta_i V_i + \delta_i \tag{4.77}$$

and

$$\eta_i = \frac{\sigma_i}{\lambda_{\max}(\Gamma_i)}, \delta_i = \frac{\sigma_i}{2} W_i^* W_i^{*\mathrm{T}} + \frac{1}{2}\varepsilon_i \tag{4.78}$$

In the last step, the control of nonaffine system is realized by Theorem 4.2.

4.2.4 Simulation Example

Example 4.3 The system is described as follows:

$$\dot{x}(t) = x^3(t) + x(t) - x(t)u(t) + 2u(t) \tag{4.79}$$

The initial condition of the system is $x(0) = 1$, and the expected system output is $y_d = \sin(0.2\ t)$.

Fig. 4.12 shows that in the neural network adaptive sliding mode, controller is charged with the output of the system and in a very short period can track the preset trajectory; Fig. 4.13 shows the output from the system controller boundedness, indicating the feasibility of the controller.

4.3 SLIDING MODE CONTROLLER DESIGN FOR NONLINEAR SYSTEMS WITH MATCHING PERTURBATIONS

4.3.1 Nonlinear Dynamic Plant Model

Consider the following nonlinear difference equation model for describing a class of general single-input and single-output discrete time dynamic plants:

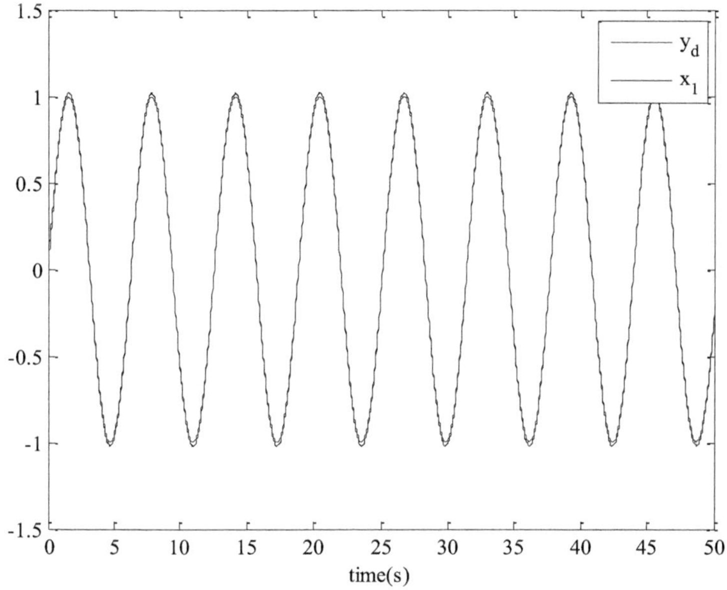

Figure 4.12 System state x_1, and prespecified trajectory $y_d = \sin(t)$.

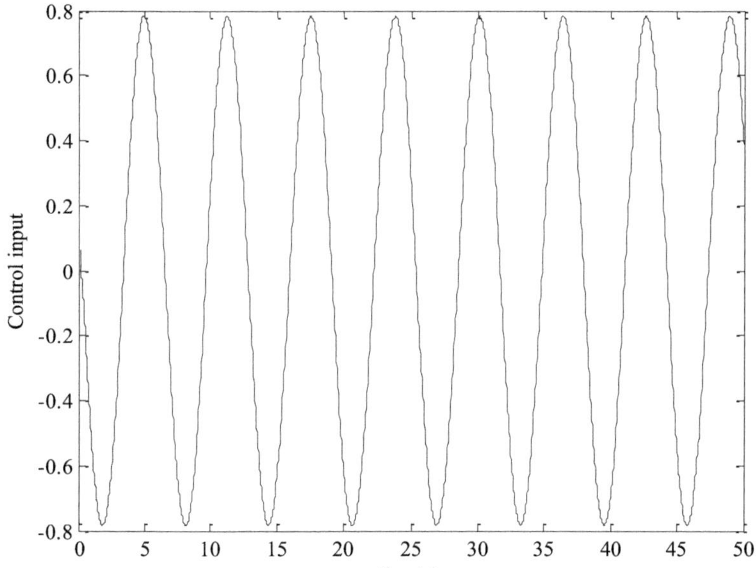

Figure 4.13 Controller output of the system.

$$\begin{aligned} y(k+1) &= f\begin{pmatrix} y(k), u(k), y(k-1) \\ , u(k-1), \cdots y(k-n), u(k-n) \end{pmatrix} \\ &== \sum_{i=0}^{L} p_i(t)\theta_i \end{aligned} \tag{4.80}$$

where $y(k) \in R$ and $u(k-1) \in R$ are the output and input (also known as controller output in control system design) signals of the plant, respectively, at discrete time instant K (1,...)n is the plant order, and $f(\bullet)$is a smooth nonlinear function. For the second line of the equation, the regression terms $p_i(k)$ are the products of past inputs and outputs such as $u(k-1)y(k-3)$,$u(k-1)u(k-2)$ $y^2(k-1)$ and are the associated parameters. Note that here the plant delay has been assumed as one for the sake of brevity but without losing generality.

4.3.2 U-Model Method

U-model methodology is a generic systematic approach to convert the nonlinear polynomial model into a controller output based on time-varying polynomial model (Zhu and Guo, 2002). U-model—enhanced control system design has been studied (Shafiq and Haseebuddin, 2005; Zhang et al., 2016) for over the past decade. Consequently, linear polynomial model—based design approaches (such as pole placement and general predictive control) can be directly used to design nonlinear control systems. The major contribution of the U-model—based design procedure can be listed in order. (1) In methodology, those well-known approaches developed from linear systems can be directly applied to nonlinear control system design, which significantly reduces the design complexity and effectively provides straightforward computational algorithms. It should be noted that this new approach does not request to linearize nonlinear plant models in design. It is just to use linear design approaches to directly design nonlinear plant model—based control systems. (2) In design, it obtains desired plant output first (compared to designing desired controller output in classical framework) and then works out the controller output from the U-model in a relaxed root-resolving routine (compared to resolving complex solutions from the inverse of the whole designed systems). (3) For linear control system design, it provides new insight and solutions within a more general and effective framework.

4.3.3 U-Model

To facilitate the control system design of such plants, the corresponding U-model, a control oriented node, has been defined by expanding the nonlinear function $f(\bullet)$ of as a polynomial with respect to $u(k-1)$ as follows:

$$y(k) = \sum_{j=0}^{M} \lambda_j(k) u^j(k-1) \tag{4.81}$$

where M is the degree of model input (controller output), the time varying parameter vector $\lambda(k) = [\lambda_0(k) \quad \cdots \quad \lambda_M(k)] \in R^{M+1}$ is a function of past inputs and outputs $u(k-2), \cdots, u(k-n)$, and the parameters $\theta_0, \cdots, \theta_L$. By this arrangement, the control-oriented model can be treated as a pure power series of input $u(k-1)$ with associated time-varying parameters $\lambda_i(k)$. Such an example is shown below:

$$\begin{aligned} y(k) &= 0.1y(k-1)y(k-2) + 0.8u(k-1)u(k-2) \\ &\quad - 0.5y(k-1)u^2(k-1) \end{aligned} \tag{4.82}$$

This can be rewritten in notation of Eq. (4.82)

$$y(k) = \lambda_0(k) + \lambda_1(k)u(k-1) + \lambda_2(k)u^2(k-1) \tag{4.83}$$

where $\lambda_0(k) = 0.1\ y(k-1)y(k-2), \lambda_1(k) = 0.8u(k-2)$, and $\lambda_2(k) = -0.5\ y(k-1)$.

4.3.4 Design of U-Model Sliding Mode Control

Consider the following nonlinear system with matching perturbations:

$$x(k+1) = Ax(k) + B\varphi(u(k)) \tag{4.84}$$

where $x(k) = [x_1(k), x_2(k), \ldots x_n(k)]^T$ is the state vector, $ABCD$ are the system parameters, respectively, and $\varphi(u(k))$is the nonlinear controller.

A discrete time system is said to satisfy the reaching condition if the following equation holds:

$$s^2(k+1) < s^2(k) \tag{4.85}$$

Choose the sliding mode as:

$$s(k) = C_e x(k) \tag{4.86}$$

Design the U-sliding controller as follow based on (Liu, 2015):

$$\varphi(u(k)) = -(C_e B)^{-1} C_e (A - I) x(k) + F_D x(k) \tag{4.87}$$

where $F_D = [f_1, \ldots f_n]$ is the nonlinear controller solved in this section, C_e is row vector, and $C_e B$ is invertible. The method has been proved by. According to Taylor series, for analytic functions of the Taylor series at a given point are finite-order functions of its Taylor's series, which completely determines the function in some neighborhood of the point. So the following polynomial equation in the current control term $u(k)$ was proposed to approximate the nonlinear controller $\varphi(u(k))$ that is described by Eq. (4.84).

$$\varphi(u(k)) = \sum_{j=0}^{M} \alpha_j(k) u^j(k-1) \tag{4.88}$$

The control input $u(k)$ can be obtained by Newton–Raphson algorithm.

4.3.5 Stability Analysis

To illustrate the effectiveness of the controller, the Lyapunov stability analysis will be discuss in the part.

First, choose a Lyapunov function as follow:

$$V(k) = \frac{1}{2} s(k)^2 \tag{4.89}$$

According to Eq. (4.86), we have:

$$s(k+1) = C_e x(k+1) = C_e A x(k) + C_e B \varphi(u(k)) \tag{4.90}$$

Substituting the controller in Eq. (4.88), we have

$$\begin{aligned} s(k+1) &= C_eAx(k) - C_e(A-I)x(k) + C_eBF_Dx(k) \\ &= C_ex(k) + C_eBF_Dx(k) \\ &= s(k) + C_eBF_Dx(k) \end{aligned} \tag{4.91}$$

So we have

$$s(k+1)^2 - s(k)^2 = 2s(k)C_eBF_Dx(k) + (C_eBF_Dx(k))^2 \tag{4.92}$$

If we need $s(k+1)^2 - s(k)^2 < 0$, it only requires $(C_eBF_Dx(k))^2 < -2s(k)C_eBF_Dx(k)$.

That is, $\frac{1}{2}(C_eB)^2\left(\sum_{i=1}^{n} f_i x_i(k)\right)^2 < -s(k)C_eB\sum_{i=1}^{n} f_i x_i(k)$.

Let $|f_i| = f_0, f_0 > 0$ for $i = 1,\ldots,n$, we have

$$\frac{1}{2}(C_eB)^2\left|\sum_{i=1}^{n} f_i x_i(k)\right|\sum_{i=1}^{n}|f_i x_i(k)| < -s(k)C_eBf_i x_i(k) \tag{4.93}$$

Let

$$\delta_i = \frac{1}{2}f_0(C_eB)^2|x_i(k)|\sum_{i=1}^{n}|x_i(k)| \tag{4.94}$$

Substituting Eq. (4.94) in Eq. (4.93), we have:

$$f_0\delta_i < -s(k)C_eBf_i x_i(k) \tag{4.95}$$

Then, we have

$$|s(k)C_eB(x_i(k))| > \delta_i \tag{4.96}$$

That is,

$$|s(k)| > \frac{\delta_i}{|C_eBx_i(k)|} \tag{4.97}$$

Substituting Eq. (4.94) in Eq. (4.97), we have:

$$|s(k)| > \frac{\frac{1}{2}f_0(C_eB)^2|x_i(k)|\sum_{i=1}^{n}|x_i(k)|}{|C_eBx_i(k)|} = \frac{f_0|C_eB|\sum_{i=1}^{n}|x_i(k)|}{2} \tag{4.98}$$

Then, we have

$$f_0 < \frac{2|s(k)|}{|C_eB|\sum_{i=1}^{n}|x_i(k)|} \tag{4.99}$$

4.3.6 Simulation Example

The example was selected to validate the performance of the proposed procedure and also to illustrate the applications. The numerical experiments were conducted with Matlab simulations.

Example 4.4 Conditional stable nonlinear process model with matching perturbations based on a sewage treatment process (Zhang, 2012) as follows:

$$\begin{aligned} x(k+1) &= Ax(k) + B\varphi(u(k)) \\ y(k+1) &= Cx(k) + D \end{aligned} \tag{4.100}$$

where

$$A = \begin{bmatrix} 3 & 0 & 1 \\ -10 & -7 & 0 \\ -5 & 0 & 1 \end{bmatrix}, \ \mathrm{B} = \begin{bmatrix} 0 \\ 0 \\ 1 \end{bmatrix}, \ \mathrm{C} = [0 \quad 0 \quad 1], \ \mathrm{D} = 0,$$

The nonlinear controller outputs are expressed in terms of Eq. (4.88):

$$[\varphi(u(k))] = [u^3(k) + u^2(k) + u(k) + 1] \tag{4.101}$$

The sliding mode is chosen as

$$s(k) = [20 \quad 10 \quad 1]\begin{bmatrix} x1 \\ x2 \\ x3 \end{bmatrix} \tag{4.102}$$

The equivalent control law is designed as

$$\varphi(u(k)) = -(C_eB)^{-1}C_e(A-I)x(k) + F_Dx(k) \tag{4.103}$$

where

$$C_e = [20 \quad 10 \quad 1] \quad F_D = [f_1 \quad f_2 \quad f_3] \quad f_0 = \frac{2|s(k)|}{|C_e B| \sum_{i=1}^{3} |x_i(k)| + 1.0}$$

Figs. 4.14–4.16 shows that the state trajectories are stable all rapidly. Fig. 4.17 shows the sliding function s also into small bounded region rapidly. Fig. 4.18 shows that the controller output trajectories obtained by Newton–Raphson algorithm.

Note that the controllers proposed in (Liu, 2015) can be used for discrete time linear system, but it does not contain controller nonlinearities. It should be noted that without U-model approach, it is not easy to design controller output $u(k)$ for nonlinear systems, while $v(k)$ is appeared in denominator polynomials. Therefore, the controller proposed in (Zhu and Guo, 2002) cannot be used directly for those systems with controller nonlinearities.

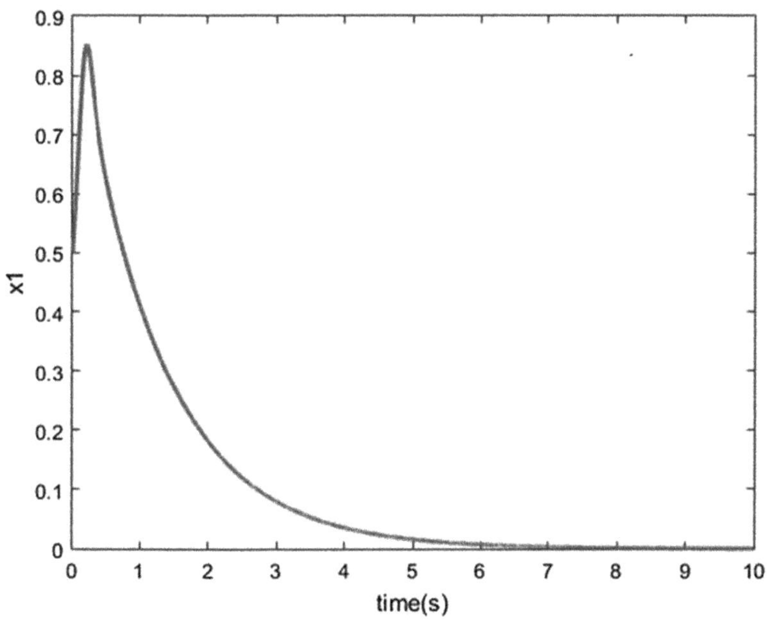

Figure 4.14 State trajectories.

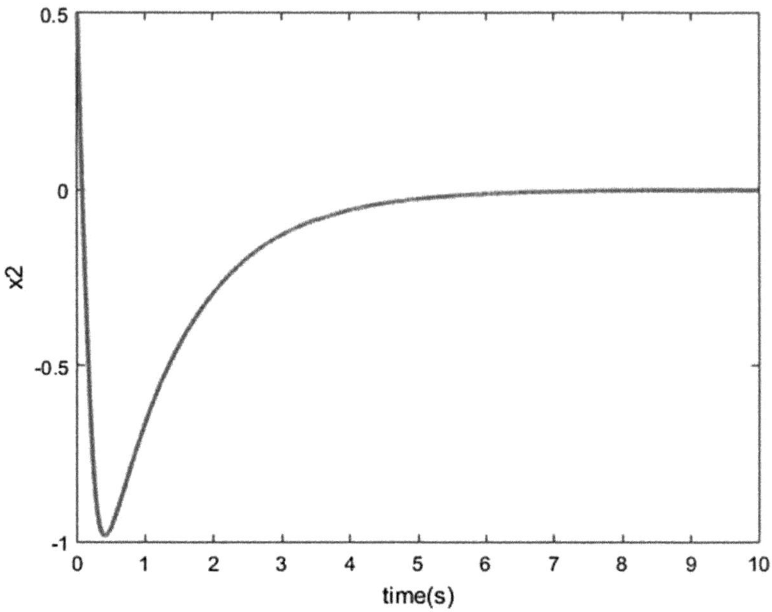

Figure 4.15 State trajectories.

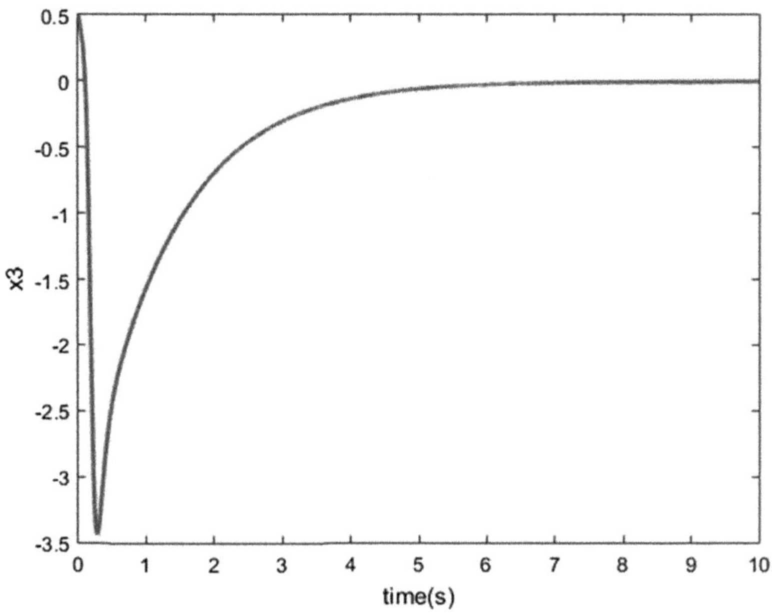

Figure 4.16 State trajectories.

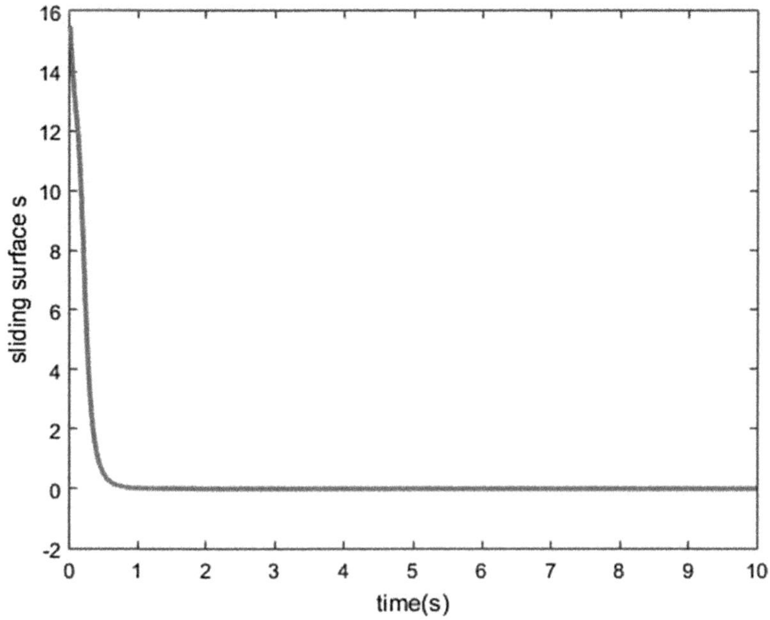

Figure 4.17 Sliding function.

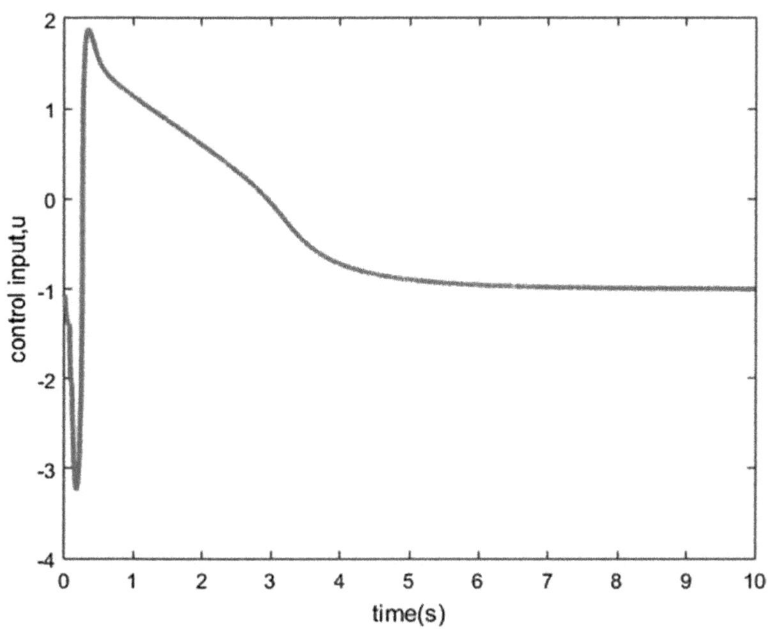

Figure 4.18 Controller input.

REFERENCES

Liu, J., 2015. Sliding Mode Control Design and MATLAB Simulation: The Basic Theory and Design Method. Tsinghua University Press.

Shafiq, M., Haseebuddin, M., 2005. U-model-based internal model control for non-linear dynamic plants. Proceedings of the Institution of Mechanical Engineers - Part I: Journal of Systems and Control Engineering 219, 449–458.

Zhang, J., Wu, X., Li, Y., Huo, J., 2016. Super Twisting sliding mode control of nonlinear system based on U model control. Control Conference 3579–3583.

Zhang, Q., 2012. The Sliding Mode Control Study on Dissolved Oxygen Concentration of Sewage Treatment Process (Dissertation for the Master Degree).

Zhu, Q.M., Guo, L.Z., 2002. A pole placement controller for non-linear dynamic plants. Acoustics Speech and Signal Processing Newsletter IEEE 216, 467–476.

CHAPTER 5
Intelligent Methodology

Contents

Chapter 5, first, studies a generalized procedure in the identification and control of a class of time-varying, delayed, nonlinear dynamic systems. Under the framework, recurrent neural network is developed to accommodate the online identification, which the weights of the neural network are iteratively and adaptively updated through the model errors. Then, indirect adaptive controller is designed based on the adaptive law of the controller, which the adaptive law of the controller is designed for the controller design. Second, it studies a neural network—based, finite-time adaptive controller for a class of uncertain, nonaffine, nonlinear systems in pure feedback prototype. In this framework, homeomorphism mapping and finite-time convergence are introduced to ensure the state constraint and convergence in finite time, respectively. To prevent constraint violation, the homeomorphism mapping is applied to a chosen compact superset, which contains initial

Adaptive Sliding Mode Neural Network Control for Nonlinear Systems
ISBN 978-0-12-815372-7
https://doi.org/10.1016/B978-0-12-815372-7.00005-7

condition and virtual superset without constraint, then use the finite-time converge method to ensure the converge time after the mapping transform. Accordingly, it provides a generic prototype for finite-time adaptive law to train neural networks.

5.1 NEURAL NETWORK IDENTIFICATION AND CONTROL FOR NONLINEAR DYNAMIC SYSTEMS

Neurocomputing techniques and algorithms have been extensively developed and applied to many different fields (Li et al., 2016). This study falls in a domain with the problem of identifying and controlling a class of complex nonlinear dynamic systems, and the solution of being developed into an integrated procedure embedded with neurocomputing, iterative learning, online identification, and adaptive control. The rest of the introduction section includes the background of the study, the state of the art of the related research, and the motivation and justification of the study, which will deliver a clear picture to the problems and solutions for the potential readers and users before going into detailed implementation of the solutions.

In the past decade, control system design for uncertain nonlinear systems (Liu et al., 2008) has received much more attention, particularly in the research of using universal function approximators to parameterize unknown nonlinearities. Neural network—based inductive techniques, subject to their inherent approximation capabilities, have been found to be very supportive for controlling such class of complex dynamic systems (Liu et al., 2008; Zhang et al., 2009). Unavoidably, online model identification is a kernel part of such control system design, and neural networks have been very effective to approximate a wide class of complex nonlinear systems in case of no full-model information achievable or even in the case of those plants with black box structures, in which the sole information is the measured input and output data sequences.

In structure, neural networks can be classified as feedforward and recurrent ones. Feedforward networks are suitable for the approximation of complex static functions. The major drawback of such type of neural networks in describing dynamic functions is that the weights' updating does not utilize information on the local date structure, and the function approximation is sensitive to the purity of training data. On the other hand, the recurrent networks incorporate feedback, not merely having concise structure but importantly, adaptive mechanism incorporated to fine-tune the approximation accuracy and convergent speed.

One of the major neural network–based adaptive control approaches is based on the Lyapunov stability theory, which gives an adaptive control law with guaranteed stability of the closed-loop systems (Jes et al., 2007; Rubio and Yu, 2007; Yu, 2004, 2006; Wu et al., 2010; Zhang et al., 2013). In particular, Polycarpou (1996) developed an adaptive neural network control scheme for uncertain strict-feedback nonlinear systems using the backstepping technique. This approach relaxed the matching condition on the plant uncertainties. The design procedure made use of linearly parameterized neural networks such as radial basis function networks with fixed centers and widths. Adaptive backstepping design was proposed for parametric strict-feedback systems with over parameterization, which could guarantee globally stable and asymptotic tracking performance. By introducing tuning functions, the over parameterization can be effectively eliminated. The adaptive backstepping design has been further extended to parametric strict-feedback systems with unknown virtual control coefficients.

A notable contribution by Ge et al. (2003, 2005) is the construction of an integral Lyapunov function, which is proved to be the key for the success of the approach. A function approximator has been used to describe the unknown nonlinear functions. Two types of artificial neural networks, the linearly parameterized neural networks (Ge et al., 2003, 2005) and the multilayer neural networks (Lewis et al., 1996; Wang and Huang, 2002; Zhang et al., 1999), have been predominantly used for approximating a wide range of unknown nonlinear functions in control system design.

Generally, all these studies referred above have presented a unified and general framework for nonlinear control system design in terms of either pure feedback or strict feedback, in which a neural network is used to approximate unknown nonlinear functions and/or uncertain terms in the systems and then based on the identified neural network's design controllers. However, from author's recent studies, there are still a number of issues that should have been further studied, and the corresponding solutions should have been developed.

5.1.1 System Description and Preliminaries

A continuous, time-varying, delay, nonlinear system can be generally described as

$$\dot{x}(t) = h(x(t), x(t-\tau(t)), u(t)) \tag{5.1}$$

where $x(t) = (x_1(t), x_2(t), \ldots, x_n(t))^T \in \Re^n$ is the state vector, $x(t-\tau(t))$ is state vector with bounded time-varying delays, $u(t) = (u_1(t), u_2(t), \ldots, u_n(t))^T \in \Re^n$ is the bound input vector, and h is an unknown continuous, nonlinear, smooth function. In this study, it is assumed that each subsystem has one-to-one structure that is only one input to act on each subsystem. Mathematically, this can be expressed as

$$\dot{x}_i(t) = h_i(x(t), x(t-\tau(t)), u_i(t))$$
$$i = 1, 2, \ldots n$$

It should be mentioned that there is no need to keep the dimensions equalized (dim(u) = dim(x)), for example, if dim(u) < dim(x) (Rubio and Yu, 2007), then the new method allows some of $u_i = 0$. Second, the general model can be used to describe nonlinear relationships between the inputs and states, instead of just linear input actions to the system states.

A delayed recurrent neural network $h_{nn}(x)$ will be used to approximate the continuous function $h(x)$, so that $h(x) = h_{nn}(x) + \widetilde{h}(x)$, where $\widetilde{h}(t)$ is the modeling error. The neural network is formatted as

$$h_{nn} = C^* x(t) + A^* f(x(t)) + B^* f(x(t - \tau(t))) + S^*$$

whereC^*, A^*, and $B^*$are the weights and S^* is the threshold of the neural network $C^* \in \Re^{n \times n}, A^* \in \Re^{n \times n}, B^* \in \Re^{n \times n}, S^* \in \Re^n$. The whole structure of the neural network is functional and concise to facilitate the utilization of Lyapunov stability theorem and adaptive laws in the following designs.

The nonlinear system (Eq. 5.1) can be written in the form of following neural networks:

$$\dot{x}(t) = C\overline{x}(t) + Af(\overline{x}(t)) + Bf(x(t - \tau(t))) + S - \widetilde{h}(t) \qquad (5.2)$$

where $\forall t \in (0, \infty)$, $C = \begin{pmatrix} C_1^T & C_2^T \end{pmatrix}^T, \overline{x}(t) = \begin{pmatrix} x^T(t) & u^T(t) \end{pmatrix}^T$, $A = \begin{pmatrix} A_1^T & A_2^T \end{pmatrix}^T$, and $C_1 \in \Re^{n \times n}$, $C_2 = diag(c_{1,n+1}, c_{2,n+2}, \ldots, c_{n,2n}) \in \Re^n$, $A_1 \in \Re^{n \times n}$, $A_2 = diag(a_{1,n+1}, a_{2,n+2}, \ldots, a_{n,2n}) \in \Re^n, B \in \Re^{n \times n}$, $S \in \Re^n$, vector $x(t)$ is the state of the neural network, $x(t - \tau(t))$ is the delayed state of the neural network, with $x(\theta) = \phi(\theta)$ $\forall \theta \in [-\tau, 0]$, $u(t)$ is the control vector, and the $\|u(t)\| \leq \overline{u}, \tau(t)$ is the time-varying delay, the ideal weights C, A, B are the bounded matrices, f is the active function, and $\widetilde{h}(t)$ is the modeling error. Because the state and output variables are physically bounded, the modeling error $\widetilde{h}(t)$ can be assumed to be bounded and $\left|\widetilde{h}(t)\right| < \varepsilon$.

For the known time-delay $\tau(t)$, it is bounded with

$$0 \leq \tau(t) \leq \tau, \quad |\dot{\tau}(t)| \leq \xi < 1$$

The active function is specified as a monotonically increased function and bounded with

$$0 \leq f(x) - f(y) \leq k(x - y)$$

For any $x, y, k \in R$ and $x \leq y, k > 0$ such as

$$f(x) = \tanh(x)$$

Selection of this kind of active function would guarantee the controller exist and obtainable by dichotomy principle.

The ideal weights C, A, B, S are defined as follows:

$$(C, A, B, S) = \arg \min_{(C,A,B,S)} \{\sup|h_{nn}(x) - h(x)|\}.$$

5.1.2 System Identification With Neural Network

The nonlinear system (Eq. 5.1) can be approximated by the following continuous-time delayed neural network:

$$\begin{aligned}\dot{y}(t) = {} & C(t)\bar{y}(t) + A(t)f(\bar{y}(t)) + B(t)f(y(t - \tau(t))) + S(t) + L(t)e(t) \\ & - diag(\text{sgn}(e(t)))^{*}\varepsilon(t)\end{aligned} \tag{5.3}$$

where $\bar{y}(t) = \begin{pmatrix} y^T(t) & u^T(t) \end{pmatrix}^T$ is the neural network output vector, that is, the estimate of the system state vector, $C(t)$,$A(t)$,$B(t)$ are interconnection weight matrices of delayed neural networks, and $S(t)$ is the threshold of neural networks. Identification error is defined as $e(t) = y(t) - x(t)$, and $L(t)$ is a diagonal constant matrix. The error derivative is obtained from, which is given below

$$\begin{aligned}\dot{e}(t) &= \dot{y}(t) - \dot{x}(t) \\ &= \widetilde{C}\bar{y}(t) + C_1 e(t) + \widetilde{A}f(\bar{y}(t)) + A_1\widetilde{f}(e(t)) + \widetilde{B}f(y(t - \tau(t))) \\ &\quad + B\widetilde{f}(e(t - \tau(t))) + \widetilde{S} + L(t)e(t) - diag(\text{sgn}(e(t)))^{*}\varepsilon(t) \\ &\quad + \widetilde{h}(t)\end{aligned} \tag{5.4}$$

where $e(t) = 0$ for $t \in [-\tau, 0]$,

$$\begin{aligned}\widetilde{C} &= C(t) - C, \widetilde{A} = A(t) - A \\ \widetilde{B} &= B(t) - B, \widetilde{S} = S(t) - S \\ \widetilde{L} &= L(t) - L, \widetilde{\varepsilon} = \varepsilon(t) - \varepsilon \\ \widetilde{f}(e(t)) &= f(y(t)) - f(x(t))\end{aligned} \tag{5.5}$$

Theorem 5.1 The identification error derivative (Eq. 5.4) is uniformly ultimately bounded (UUB), where the weights of the delayed neural network are updated through following equations.

$$
\begin{aligned}
\dot{C}(t) &= -\Gamma_1\big(e(t)\overline{y}^T(t) + \sigma_1 C(t)\big) \\
\dot{A}(t) &= -\Gamma_2\big(e(t)f^T(\overline{y}(t)) + \sigma_2 A(t)\big) \\
\dot{B}(t) &= -\Gamma_3(e(t)f^T(y(t-\tau(t))) + \sigma_3 B(t)) \\
\dot{S}(t) &= -\Gamma_4(e(t) + \sigma_4 S(t)) \\
\dot{L}(t) &= -\Gamma_5(e(t)e^T(t) + \sigma_5 L(t)) \\
\dot{\varepsilon}(t) &= -\Gamma_6(-|e(t)| + \sigma_6 \varepsilon(t))
\end{aligned} \tag{5.6}
$$

where Γ_i is a positive definite matrix and σ_i is a constant, such that $\Gamma_i = \Gamma_i^T > 0,\ \sigma_i > 0,\ i = 1,2\ldots\ 6$.

Proof Select a Lyapunov–Krasovskii candidate function as

$$
V_1 = \frac{1}{2}e^T(t)e(t) \tag{5.7}
$$

The derivative is as follow

$$
\begin{aligned}
\dot{V}_1 = e^T(t)\widetilde{C}\overline{y}(t) + e^T(t)C_1e(t) + e^T(t)\widetilde{A}f(\overline{y}(t)) + e^T(t)A_1\widetilde{f}(e(t)) \\
+e^T(t)\widetilde{B}f(y(t-\tau(t))) + e^T(t)B\widetilde{f}(e(t-\tau(t))) \\
+e^T(t)\widetilde{S} + e^T(t)L(t)e(t) - |e^T(t)|\varepsilon(t) + e^T(t)\widetilde{h}(t)
\end{aligned} \tag{5.8}
$$

With taking the active function into consideration, it gives

$$
\begin{aligned}
e^T(t)A_1\widetilde{f}(e(t)) &\le \frac{1}{2}e^T(t)A_1A_1^Te(t) + \frac{k^2}{2}e^T(t)e(t) \\
e^T(t)B\widetilde{f}(e(t-\tau(t))) &\le \frac{1}{2}e^T(t)BB^Te(t) + \frac{k^2}{2}e^T(t-\tau(t))e(t-\tau(t))
\end{aligned} \tag{5.9}
$$

Accordingly, the following inequality is held

$$
\begin{aligned}
\dot{V}_1 \le -\eta_1 V_1 &+ e^T(t)\widetilde{C}\overline{y}(t) + e^T(t)\widetilde{A}f(\overline{y}(t)) + e^T(t)\widetilde{B}f(y(t-\tau(t))) \\
&+ e^T(t)\widetilde{S} + e^T(t)\widetilde{L}e(t) - |e^T(t)|\widetilde{\varepsilon}(t) \\
&+ \frac{k^2}{2}e^T(t-\tau(t))e(t-\tau(t)) + e^T(t)\bigg[C_1 + \frac{1}{2}A_1A_1^T + \frac{1}{2}BB^T \\
&+ L + \frac{k^2+2}{2}\bigg]e(t)
\end{aligned} \tag{5.10}
$$

where $\eta_1 = 1$ because $\widetilde{C}, \widetilde{A}, \widetilde{B}, \widetilde{S}$ contain the ideal weight matrices, so cancelled them in the next step.

$$V_2 = \frac{1}{2} tr\left(\widetilde{C}^T \Gamma_1^{-1} \widetilde{C} + \widetilde{A}^T \Gamma_2^{-1} \widetilde{A} + \widetilde{B}^T \Gamma_3^{-1} \widetilde{B} + \widetilde{S}^T \Gamma_4^{-1} \widetilde{S} + \widetilde{L}^T \Gamma_5^{-1} \widetilde{L} + \widetilde{\varepsilon}^T \Gamma_6^{-1} \widetilde{\varepsilon}\right) \tag{5.11}$$

Accordingly, its derivative is derived as below

$$\begin{aligned}\dot{V}_2 \le & -e^T(t)\widetilde{C}\overline{y}(t) - e^T(t)\widetilde{A}f(\overline{y}(t)) - e^T(t)\widetilde{B}g(y(t-\tau(t))) - e^T(t)\widetilde{S} \\ & -e^T(t)\widetilde{L}e(t) + |e(t)|\widetilde{\varepsilon}(t) - \eta_2 V_2 + \frac{\sigma_1}{2} tr\left(C_1^T C_1\right) + \frac{\sigma_2}{2} tr\left(A_1^T A_1\right) \\ & + \frac{\sigma_3}{2} tr(B^T B) + \frac{\sigma_4}{2} tr(S^T S) + \frac{\sigma_5}{2} tr(L^T L) + \frac{\sigma_6}{2} tr(\varepsilon^T \varepsilon)\end{aligned} \tag{5.12}$$

where $\eta_2 = \max\left(\frac{\sigma_i}{\lambda_{max}\left(\Gamma_i^{-1}\right)}\right), i = 1, 2, \ldots, 6.$

Assign V_3 to cancel the sections associated with time delay in $\dot{V}_1$.

$$V_3 = \frac{k^2}{2(1-\xi)} \int_{t-\tau(t)}^{t} \exp\left(\frac{t-\tau(t)-\theta}{\xi-1}\right) e^T(\theta)e(\theta)d\theta \tag{5.13}$$

and its derivative is

$$\begin{aligned}\dot{V}_3 \le & -\eta_3 V_3 + \frac{k^2}{2(1-\xi)} \exp\left(\frac{\tau}{1-\xi}\right) e^T(t)e(t) \\ & - \frac{k^2}{2} e^T(t-\tau(t))e(t-\tau(t))\end{aligned} \tag{5.14}$$

where $\eta_3 = \frac{1+\xi}{1-\xi}$. Above all, choose the total Lyapunov function as

$$V = V_1 + V_2 + V_3$$

and its derivative is

$$\dot{V} \le -\eta V + e^T(t)\Xi e(t) + \delta$$

where

$$\Xi = C_1 + \frac{1}{2}A_1A_1^T + \frac{1}{2}BB^T + L + \frac{k^2+1}{2} + \frac{k^2}{2(1-\xi)}\exp\left(\frac{\tau}{1-\xi}\right) \tag{5.15}$$

$$\begin{aligned}\delta &= \frac{\sigma_1}{2}tr\left(C_1^TC_1\right) + \frac{\sigma_2}{2}tr\left(A_1^TA_1\right) + \frac{\sigma_3}{2}tr(B^TB) + \frac{\sigma_4}{2}tr(S^TS) \\ &\quad + \frac{\sigma_5}{2}tr(L^TL) + \frac{\sigma_6}{2}tr(\varepsilon^T\varepsilon)\eta \\ &= \min(\eta_1, \eta_2, \eta_3)\end{aligned}$$

To make $\Xi < 0$ to specify L as follows

$$L < -\left(C_1 + \frac{1}{2}A_1A_1^T + \frac{1}{2}BB^T + \frac{k^2+2}{2} + \frac{k^2}{2(1-\xi)}\exp\left(\frac{\tau}{1-\xi}\right)\right)$$

δ is bound from the characteristics provided from the ideal matrices and the modeling error.

Therefore, it can be concluded that

$$\begin{aligned}&\dot{V} \le -\eta V + \delta \\ &0 \le V \le \left(V(0) - \frac{\delta}{\eta}\right)\exp(-\eta t) + \frac{\delta}{\eta} \\ &\|e(t)\| \le \sqrt{2\left(V(0) - \frac{\delta}{\eta}\right)\exp(-\eta t) + \frac{2\delta}{\eta}} \\ &\lim_{t\to\infty}\|e(t)\| = \sqrt{\frac{2\delta}{\eta}}\end{aligned}$$

With reference to boundness theorem (Ge et al., 2003), the error induced in using the neural network to approximate this class of systems is UUB. The boundary $\sqrt{\frac{2\delta}{\eta}}$ is a function of the ideal weights of the neural network.

Remark 1: In theory, the gains Gamma and Sigma could be chosen arbitrarily within their closure sets. But limited by computation algorithms, a proper selection of Gamma and Sigma is necessary for the identification and control. Similar work can be found in some articles, such as that of Ge et al. (2003).

5.1.3 Indirect Adaptive Controller Design

The aim of the controller design is to drive the outputs (states) of the system properly following a prespecified trajectory. Without losing generality, let the desired trajectory $x^*(t)$ be smooth. To describe the states of the system following the desired trajectory, a nonlinear system in category of Eq. (5.1) is presented below

$$\dot{x}(t) = h_{nn} - \widetilde{h}(t) \tag{5.16}$$

The parameters of neural network h_{nn} could be gained through the identification algorithms.

The desired trajectory is notated as $x^*(t)$, through coordinate transformation

$$z(t) = x(t) - x^*(t)$$

Accordingly, nonlinear system (Eq. 5.16) can be alternatively expressed as

$$\dot{z}(t) = h_{nn} - \widetilde{h}(t) - \dot{x}^*(t)$$

Because $x^*(t)$ is the prespecified desired trajectory either by customers or designers, the objective is to drive the output of the plant $x(t)$ following the desired trajectory $x^*(t)$.

$$h_{nn} = -k^* z(t) - \varepsilon^T(t)\text{sgn}(z(t)) + \dot{x}^*(t), k^* > 0$$

$$\dot{z}(t) = -k^* z(t) - \varepsilon^T(t)\text{sgn}(z(t)) - \widetilde{h}(t)$$

$$V = \frac{1}{2} z^T(t) z(t)$$

$$\begin{aligned} \dot{V} &= -k^* z^T(t) z(t) - |z^T(t)|\varepsilon(t) + z^T(t)\widetilde{h}(t) \\ &\leq -k^* z^T(t) z(t) \end{aligned}$$

Till this stage, the controller design task can be clearly stated as follows: Use the identified neural network (model) to represent the system behavior, with this reference model to design the control system; therefore, the controlled system output (states) can be properly driven following a prespecified desired trajectory.

If the dimension of the controller $dim(u(t)) = n$, based on the identification method

$$\begin{aligned} h_{nn} &= -k^{*}z(t) - \varepsilon(t)\text{sgn}(z(t)) + \dot{x}^{*}(t) \\ &= C_1(t)x(t) + C_2(t)u(t) + A_1(t)f(x(t)) \\ &+A_2f(u(t)) + B(t)f(x(t-\tau(t))) + S(t) \end{aligned}$$

It is requested to resolve the root of the nonlinear equations to obtain the controller output,

$$\xi\left(u^{*}\right) = D_1 + D_2u^{*}(t) + D_3f\left(u^{*}(t)\right) = 0$$

where the parameters of the equation are specified by

$$\begin{aligned} D_1 &= k^{*}z(t) + \varepsilon(t)\text{sgn}(z(t)) - \dot{x}^{*}(t) + C_1(t)x(t) + A_1(t)f(x(t)) \\ &\quad + B(t)f(x(t-\tau(t))) + S(t) \\ D_2 &= C_2(t) \\ D_3 &= A_2(t) \end{aligned}$$

Because the active function is bounded, it can be proved that the ideal controller is exist and can be resolved by dichotomy principle. To further explain the assertion, take off one of the nonlinear equations and say

$$\xi(u) = d_1 + d_2u + d_3f(u) \tag{5.17}$$

If $d_1 = 0$, then $u = 0$ is one of the roots for nonlinear Eq. (5.17);

If $d_1 \neq 0$ without losing generality, assume $d_1 > 0$, it would not affect the root of the nonlinear Eq. (5.17), then assume $d_2 \neq 0$, set up $a = 0$, $b = -\frac{\alpha(d_1+abs(d_3))}{d_2}$, where $\alpha > 1$, then $\xi(a) > 0$ and $\xi(b) < 0$. Base on the theorem of the numerical analysis, there exists an ideal controller output u^* between a and *b*. Then, $\xi(u) = 0$ can be resolved through the dichotomy principle, which is presented in appendix.

If the $d_1 \neq 0, d_2 = 0$, then the $u = f^{-1}\left(-\frac{d_1}{d_3}\right)$ when the $\left|\frac{d_1}{d_3}\right| \leq 1$, else the root of the nonlinear Eq. (5.17) does not exist.

The whole online computational algorithm for the system identification and controller design will be described in next section.

In summary, the indirect adaptive control addressed in this study is to guarantee that all the signals in the closed-loop system remain bounded, and error system state z converges to origin. From Lyapunov stability theorem, system is stable; because the error of the system is ultimately uniformly bounded, the system state x would converge to a small neighborhood around desired trajectory x^*.

5.1.4 Algorithm for Implement

In this section, a step-by-step procedure is listed to implement the control scheme.

Assign initial values of the gain matrices in adaptive law Γ, σ, L, the initial value of to be estimated parameters and variable states of $C(1), A(1), B(1), S(1), L(1), \varepsilon(1), \gamma(1)$.

Based on the initials of the state $x(1)$ and system input $u(1)$, calculate the state of the neural networks $\gamma(2)$, start the algorithm at time instant $i = 1$.

Calculate the new parameters and variables $C(i+1)$, $A(i+1)$, $B(i+1)$, $S(i+1)$ by adaptive law (Eq. 5.6) and determine the variable $\gamma(i+1)$ according to Eq. (5.3).

Assign the threshold $\varepsilon_0 = 0.01$ and system input $u(i+1) = 0$, if $|d_1| \leq \varepsilon_0$, go to Step 9.

If $|d_2| \leq \varepsilon_0$, $|d_3| > \varepsilon_0$ and $\left|\frac{d_1}{d_3}\right| \leq 1$, then $u(i+1) = f^{-1}\left(-\frac{d_1}{d_3}\right)$, go to Step 9.

If $|d_2| > \varepsilon_0$, then $d_2 = d_2{}^*\mathrm{sgn}(d_1), d_3 = d_3{}^*\mathrm{sgn}(d_1)$, and $d_1 = abs(d_1)$, assign the initial values $a = 0$, $b = -\frac{\alpha(d_1 + abs(d_3))}{d_2}$,

$\alpha = 1.1$, for dichotomy principle. Set up iterative learning step index $j = 1$.

Assign $c = 0.5(a + b)$ and then calculate $\xi(c)$ from Eq. (5.17). If $|\xi(c)| < \varepsilon_0$, then $u(i+1) = c$, go to Step 9.

If $\xi(c) > 0$, then $a = c$, else $b = c$, $\mathrm{j} = \mathrm{j}+1$, if $\mathrm{j} < 3000$ go to Step 7.

$i = i+1$ go to Step 3.

This is the online identification and control algorithm for nonlinear system with time delay. It should be mentioned that the algorithm can be solely used for the identification in either open loop or closed loop.

5.1.5 Simulation Studies

Two examples were selected to validate the performance of the proposed procedure and also to illustrate how to use the procedure in design.

Example 5.1 The first system was given by

$$\dot{x}_1 = x_1(t) - x_1(t)^* x_2(t) - x_1(t - \tau(t))^* x_2(t - \tau(t)) + u_1$$
$$\dot{x}_2 = x_1(t) + x_2(t) + x_2(t)^* u_2 + u_2$$

The system was initially located at $x(0) = (1,1)^T$. Assumed the structure of the two inputs and two outputs system unknown that is a black box system except external input-output data available. In this experiment, the first test was the feasibility to use the proposed neural network to approximate the dynamic system, which was treated as a model identification case study. Initially, set input $u_1(t) = \sin(t), u_2(t) = \cos(t)$ to identify the free response model. The cellular neural network was structured with 2 layers, 10 neurons for input layer, 2 neurons for output layer, and the activation functions were selected as hyperbolic tangent

$$\tanh(x) = \frac{e^x - e^{-x}}{e^x + e^{-x}}$$

To design the controller, it was requested to find a model to approximate the system; therefore, selected a recurrent neural network (model) as follows

$$\dot{y}(t) = C(t)\overline{y}(t) + A(t)f(\overline{y}(t)) + S(t) + L(t)e(t) \\ - diag(\text{sgn}(e(t)))^{*}\varepsilon(t)$$

which was composed with 2 layers, 8 neurons for input layer, and 2 neurons for output layer; activation functions were chosen as $\tanh(x) = \frac{e^x - e^{-x}}{e^x + e^{-x}}$. Assigned the initials $C(0) = \begin{pmatrix} 1 & 1 & 1 & 0 \\ 1 & 1 & 0 & 1 \end{pmatrix}$, $A(0), S(0), L(0), \varepsilon(0), y(0)$ as null matrices and null vectors and $\Gamma = \begin{bmatrix} 1 & 0.01 \\ 0.01 & 1 \end{bmatrix}, \sigma = 0.001$.

Then, run the online identification and controller design procedure listed in the last section. The input and output were sampled every 0.01 s and the whole process was run for 50 s.

The control task was to drive the system state $x(t) = (x_1(t), x_2(t))^T$ to follow a prespecified trajectory vector $x^*(t) = (\sin(t), \cos(t))^T$ with proper dynamic and static characteristics.

The simulated results are shown in Figs. 5.2—5.5 to indicate the excellent approximation of using the neural network for the identification and control, which the errors were quickly converged to zero. Furthermore, it can be observed with the excellent performance of the system states following the prespecified trajectory.

Example 5.2 The complex nonlinear system was described by

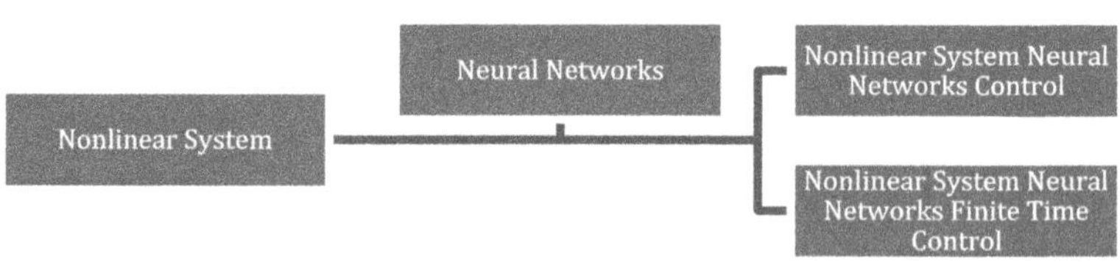

Figure 5.1 The structure chart of Chapter 5.

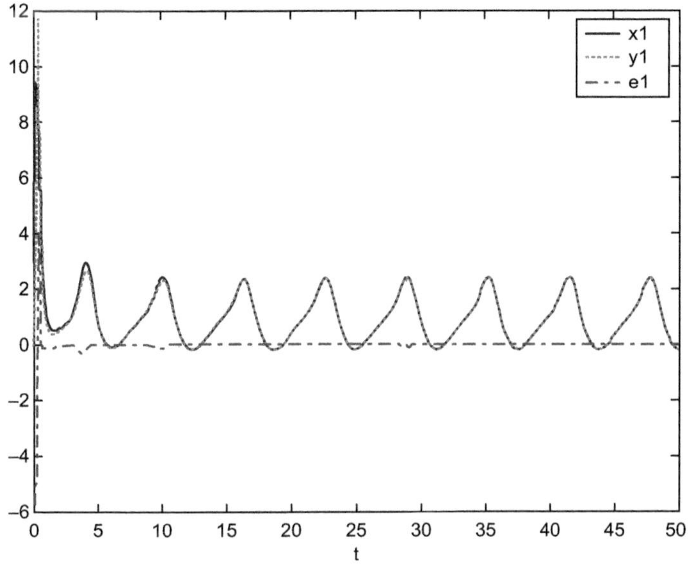

Figure 5.2 System state x_1, NN state y_1, and error state e_1 with sinusoidal input.

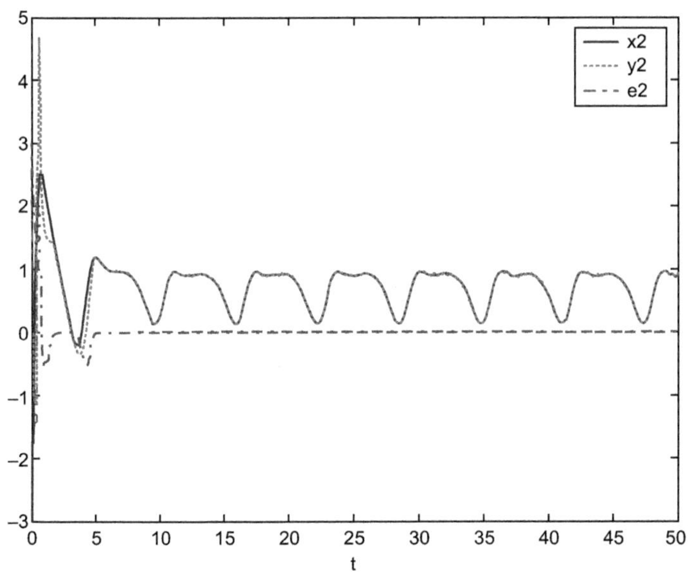

Figure 5.3 System state x_2, NN state y_2, and error state e_2 with cosine input.

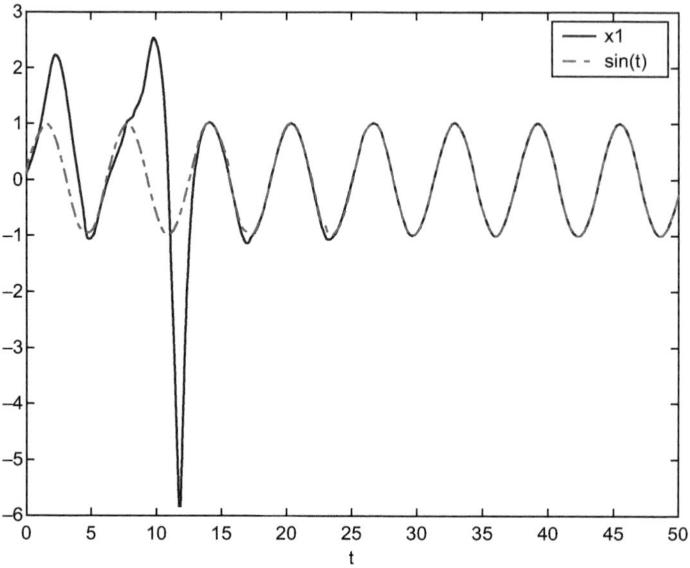

Figure 5.4 System state x_1 and prespecified trajectory $x_1^* = \sin(t)$.

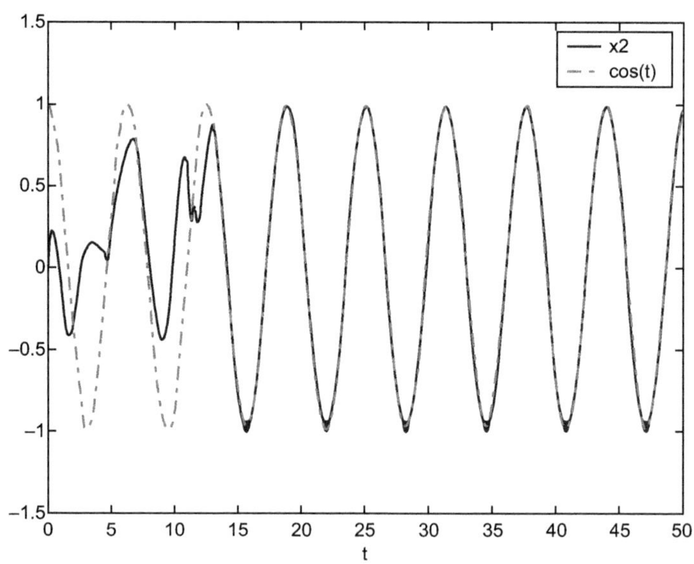

Figure 5.5 System state x_2 and prespecified trajectory $x_2^* = \cos(t)$.

$$\dot{x}_1 = x_1(t) - x_1(t)^* x_2(t) - x_1(t-\tau(t))^* x_2(t-\tau(t))$$
$$\dot{x}_2 = x_1(t) + x_2(t) + x_2(t)^* u_2 + u_2$$

Obviously, the input has implicit relationship with state $x_2(t)$ that is the input, which cannot be explicitly separated from state $x_2(t)$. The recurrent neural network was structured with 2 layers, 10 neurons for input layer, 2 neurons for output layer, and the activation functions were hyperbolic tangent, $\tanh(x) = \frac{e^x - e^{-x}}{e^x + e^{-x}}$. Assigned the initials $\Gamma = 0.1^* I, \sigma = 1$, $C(0) = \begin{pmatrix} 1 & 1 & 1 & 0 \\ 1 & 1 & 0 & 1 \end{pmatrix}$, $A(0), B(0), S(0)$ as null matrices and null vectors, initial conditions of plant were set $(x_1(t), x_2(t))^T = (3, 3)^T$, respectively. Then, run the online identification and controller design procedure listed in last section. The input and output were sampled every 0.01 s, and the whole process was run for 30 s.

Once again as demonstrated in Example 5.1, the second case study is shown in Figs. 5.6–5.9 to indicate the excellent

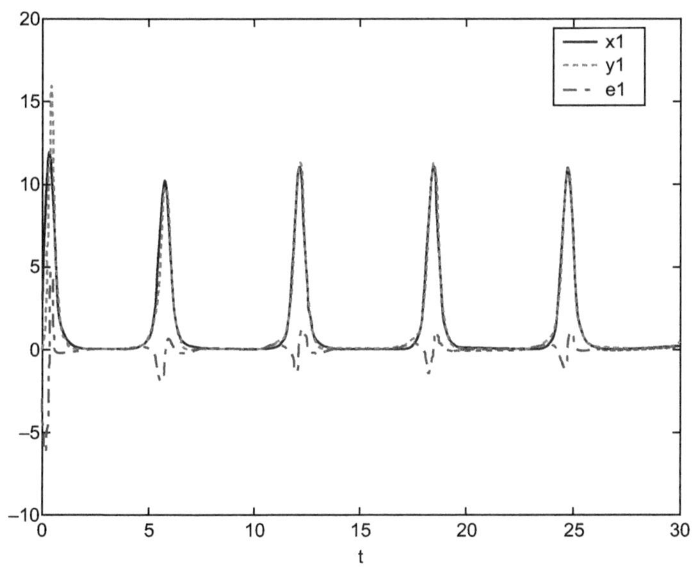

Figure 5.6 System state x_1, NN state y_1, and error state e_1 with sinusoidal input.

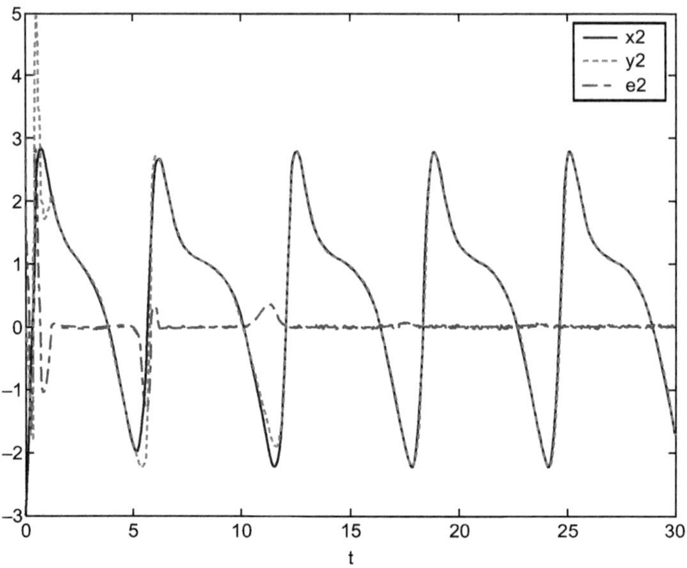

Figure 5.7 System state x_2, NN state y_2, and error state e_2 with sinusoidal input.

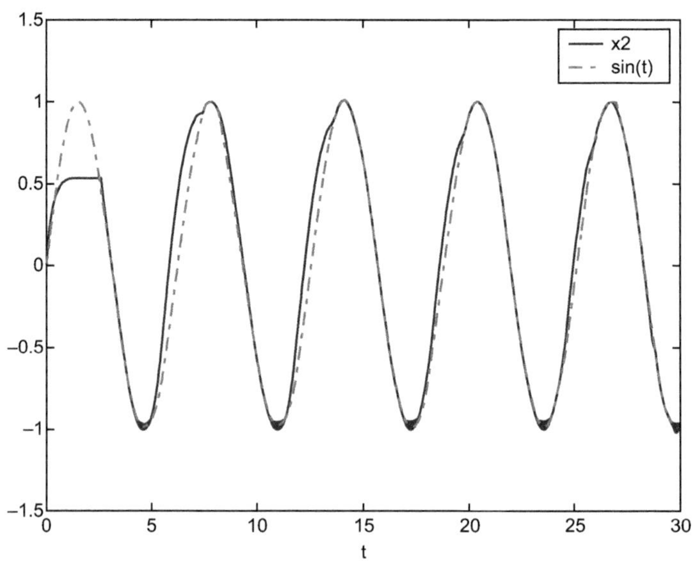

Figure 5.8 System state x_2 and prespecified trajectory $x_2^* = \sin(t)$.

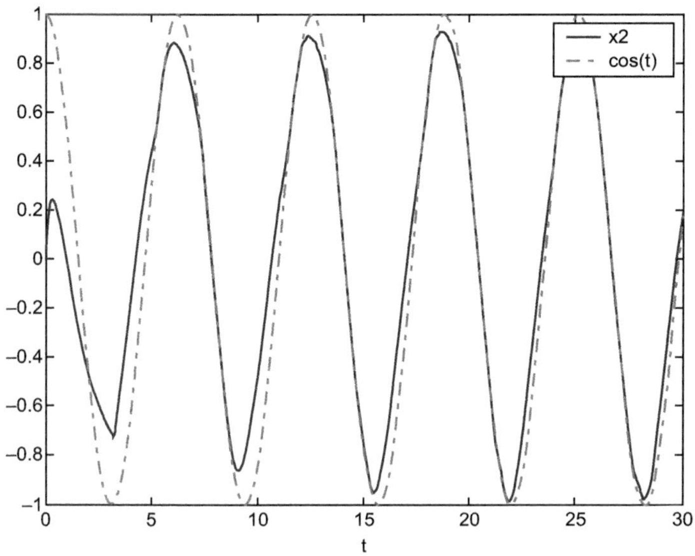

Figure 5.9 System state x_2 and prespecified trajectory $x_2^* = \cos(t)$.

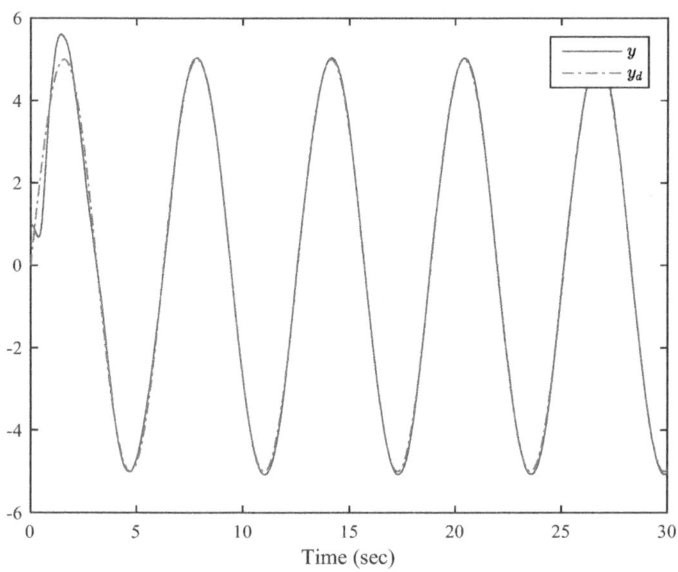

Figure 5.10 Trajectories of y, y_d.

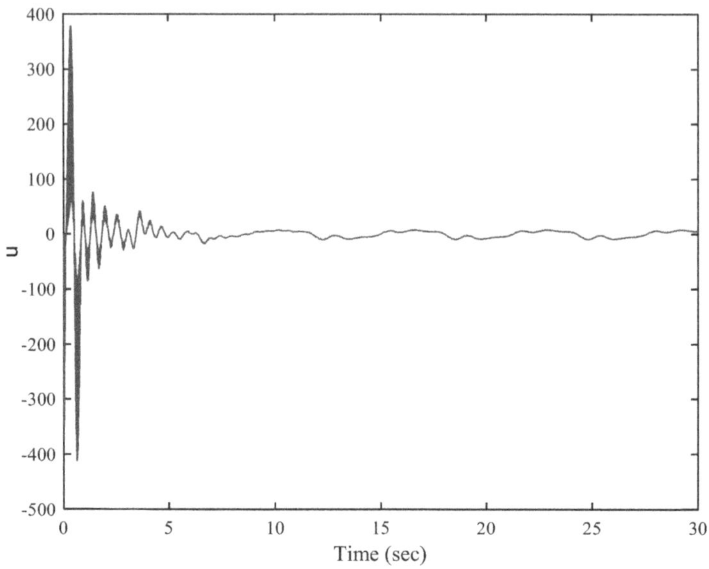

Figure 5.11 Trajectory of control input u.

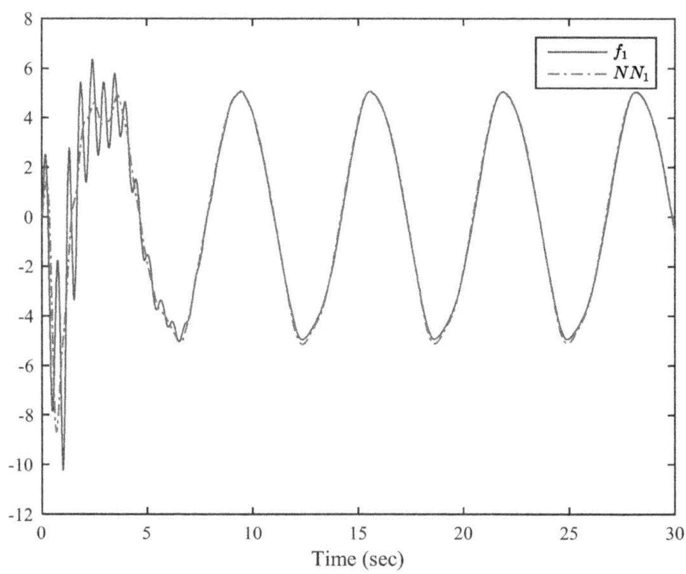

Figure 5.12 Trajectories of f_1,NN_1.

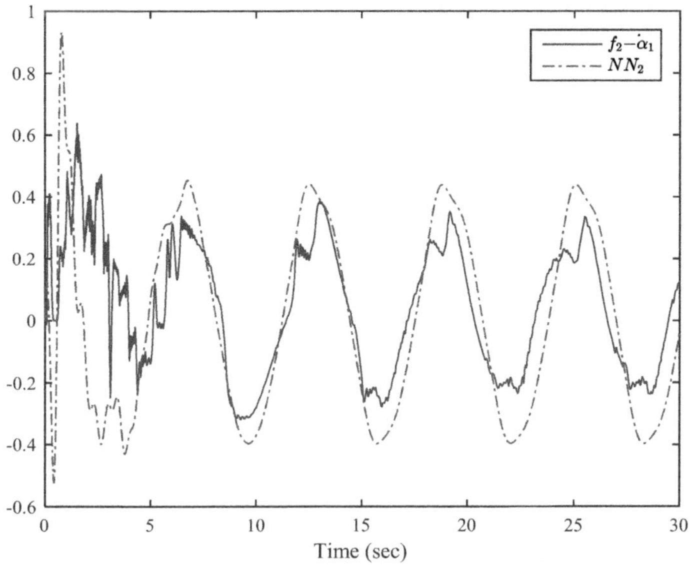

Figure 5.13 Trajectories of $f_2 - \dot{\alpha}_1$, NN_2.

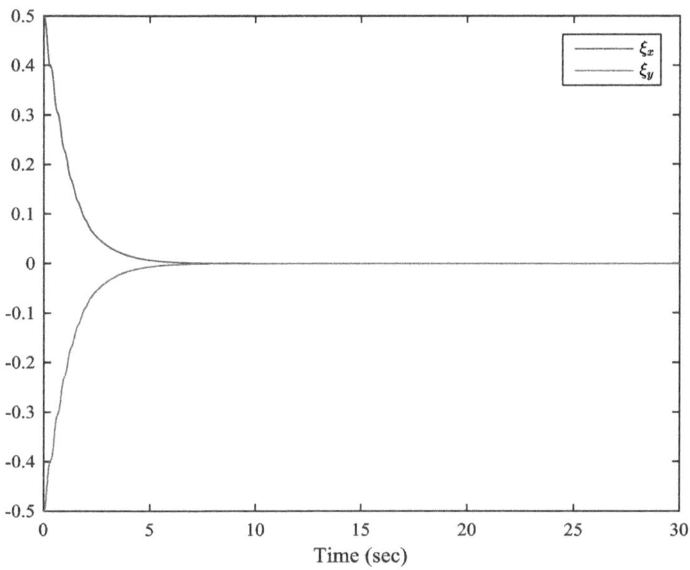

Figure 5.14 Trajectories of ξ_x, ξ_y.

approximation of using the neural network for the identification and control, which the errors were quickly converged to zero. Furthermore, it can be observed with the excellent performance of the system states following the prespecified trajectory.

5.2 FINITE-TIME ADAPTIVE NEURAL NETWORK CONTROL

5.2.1 System Description and Preliminaries

Throughout this article, the rational number η and the matrix $W \in \Re^{m\times n}$, the matrix W^η denotes element-by-element powers, where

$$W = \begin{bmatrix} w_{11} & w_{12} & \cdots & w_{1n} \\ w_{21} & w_{22} & \cdots & w_{2n} \\ \vdots & \vdots & \ddots & \vdots \\ w_{m1} & w_{m2} & \cdots & w_{mn} \end{bmatrix}, W^\eta = \begin{bmatrix} w_{11}^\eta & w_{12}^\eta & \cdots & w_{1n}^\eta \\ w_{21}^\eta & w_{22}^\eta & \cdots & w_{2n}^\eta \\ \vdots & \vdots & \ddots & \vdots \\ w_{m1}^\eta & w_{m2}^\eta & \cdots & w_{mn}^\eta \end{bmatrix} \tag{5.18}$$

$\widetilde{W}^{\eta T}$ indicates the transposition of the matrix $\widetilde{W}^{\eta}$, where

$$\widetilde{W}^{\eta T} = \left(\widetilde{W}^{\eta}\right)^T \tag{5.19}$$

A continuous nonlinear system can be generally described as following:

$$\begin{aligned} \dot{x}_i &= f_i(\overline{x}_i) + x_{i+1} \\ 1 &\leq i \leq n-1 \\ \dot{x}_n &= f_n(\overline{x}_n, u) \\ y &= x_1 \end{aligned} \tag{5.20}$$

where $\overline{x}_i = \begin{pmatrix} x_1 & x_2 & \cdots & x_i \end{pmatrix}^T \in \Re^i$, $u \in \Re$ y indicate the state, control, output, and initial condition, respectively; $f_i(\bullet)$ is nonlinear smooth functions(Shengli Shi and Jiayu, 2018).

In brief, the following radial basis function (RBF) Neural Networks (NN) (Rui Peng et al., 2018) is used to approximate the continuous function $F(x) : \Re^n \to \Re$ over a compact set

$$F_{NN}(x, W) = W^T \Psi(x) \tag{5.21}$$

where the input $x \in \Omega \subset \Re^n$, the weight vector $W = [\, w_1 \quad \cdots \quad w_l \,]^T \in \Re^l$, and the node vector $\Psi(x) = [\, \psi_1(x) \quad \cdots \quad \psi_l(x) \,]^T$, which the element $\psi_i(x)$ being chosen as the commonly used Gaussian function as below

$$\psi_i(x) = \exp\left[\frac{-(x-\mu_i)^T(x-\mu_i)}{\eta_i^2}\right], i = 1, 2, \cdots l. \tag{5.22}$$

where $\mu_i = [\, \mu_{i1} \quad \cdots \quad \mu_{in} \,]^T$, the center of the receptive field, and η_i is the width of the Gaussian function.

It has been proven that RBF NN can approximate any continuous function over a compact set $\Omega_x \subset \Re^n$ as

$$F(x) = W^{*T} \Psi(x) + \varepsilon(x) \tag{5.23}$$

where the ideal NN weight is W^*, and $\varepsilon(x)$ is the NN approximation error.

$$W^* = \arg\min_{W \in \Re^l} \{\sup |F(x) - W^T \Psi(x)|\} \tag{5.24}$$

Given a nonlinear system

$$\dot{x} = f(x, t), y = h(x, t), x \in \Re^n, t \geq t_0 \tag{5.25}$$

if there exists a compact set $U \in \Re^n$ such that for all $x(t_0) = x_0 \in U$, there exists a $\delta > 0$ and a number $T(\delta, x_0)$ such that $\|x(t)\| < \delta$ for all $t \geq t_0 + T(\delta, x_0)$, the solution is UUB (Yafei Liu et al., 2018).

Consider a system

$$\dot{x} = f(x, t), f(0, t) = 0, x \in U_0 \subset \Re^n \tag{5.26}$$

where $f: U_0 \times \Re^+ \to \Re^n$ is continuous with respect to x on an open neighborhood U_0 of the origin $x = 0$. The equilibrium $x = 0$ of the system is (locally) finite-time stable if it is Lyapunov stable and finite-time convergent in a neighborhood $U \subseteq U_0$ of the origin. By "finite-time convergence," we mean if, for any initial condition $x(t_0) = x_0 \in U$ at any given initial time t_0, there is a settling time $T > 0$, such that every solution $x(t; t_0, x_0)$ of system is defined with $x(t; t_0, x_0) \in U/\{0\}$ for $t \in [t_0, T)$ and

$$\lim_{t \to T} x(t; t_0, x_0) = 0 \quad x(t; t_0, x_0) = 0 \quad \forall t > T \tag{5.27}$$

When $U = \Re^n$, the origin is a globally finite-time stable equilibrium.

The solution is semiglobally UUB provided that for any compact set $\Omega_0 \subset \Re^n$, and all $\overline{x}_n(t_0) \in \Omega_0$, there has $\mu>0$, $T(\mu, \overline{x}_n(t_0))$ such that $\|\overline{x}_n(t)\| \leq \mu$ for all $t > t_0 + T$.

Consider the system with uniqueness of the solution in forward time. If there exist real numbers $\beta > 1$, $k > 0$, a Class-K function σ and C^1 Lyapunov function, $V(x)$ of the system such that $\sigma(\|x\|) \leq V(x)$, and $\dot{V} \leq kV^{\frac{1}{\beta}}(x)$, $x \in \Omega$ hold along the trajectory of the system whenever $V(x(t+\theta)) \leq V(x(t))$ for $\theta \in [-h, 0]$, then the system is finite-time stable.

If $\Omega = \Re^n$ and σ is a Class-K function, then the origin is a globally finite-time stable equilibrium of the system. Furthermore, the settling time of the system with respect to the initial condition $\Phi \in C_\delta$ satisfies

$$T_0(\Phi) \leq \frac{\beta}{k(\beta - 1)} V^{\frac{\beta-1}{\beta}}(\Phi) \tag{5.28}$$

for all $t \geq 0$.

For any constant $a, b \in \Re$, the following inequality holds:

$$ab \leq \frac{1}{p}a^p + \frac{1}{q}b^q \tag{5.29}$$

where $p > 1$, $q > 1$, and $\frac{1}{p} + \frac{1}{q} = 1$.

For any constant $a, b \in \Re$, the following inequality holds:

$$-a(a+b)^{\frac{1}{3}} \leq -\frac{1}{2}a^{\frac{4}{3}} + b^{\frac{4}{3}} \tag{5.30}$$

$$-\widetilde{W}^T \widehat{W}^{\frac{1}{3}} \leq -\frac{1}{2}\widetilde{W}^{\frac{2}{3T}} \widetilde{W}^{\frac{2}{3}} + W^{*\frac{2}{3T}} W^{*\frac{2}{3}} \tag{5.31}$$

$$z\varepsilon \leq \frac{3}{4}z^{\frac{4}{3}} + \frac{1}{4}\varepsilon^4 \tag{5.32}$$

Proof: Based on the Young's inequality, the following inequality is hold

$$\begin{aligned} -a^3 b &\le \frac{3}{4}(a^3)^{\frac{4}{3}} + \frac{1}{4}b^4 \\ &= \frac{3}{4}a^4 + \frac{1}{4}b^4 \end{aligned} \tag{5.33}$$

and

$$\frac{3}{2}a^{\frac{4}{3}}b^{\frac{8}{3}} = \frac{3}{2} \times a^{\frac{4}{3}}b^{\frac{2}{3}} \times b^2 \le \frac{3}{2}\left(\frac{\left(a^{\frac{4}{3}}b^{\frac{2}{3}}\right)^2 + b^4}{2}\right) \le \frac{3}{4}a^{\frac{8}{3}}b^{\frac{4}{3}} + \frac{3}{4}b^4 \tag{5.34}$$

Based on inequality Eqs. (5.33) and (5.34), we have

$$-a^4 - a^3 b + \frac{3}{2}a^{\frac{4}{3}}b^{\frac{8}{3}} \le -\frac{1}{4}a^4 + b^4 + \frac{3}{4}a^{\frac{8}{3}}b^{\frac{4}{3}} \tag{5.35}$$

Then,

$$-a^4 - a^3 b \le -\frac{1}{8}a^4 + b^4 + \frac{3}{4}a^{\frac{8}{3}}b^{\frac{4}{3}} - \frac{3}{2}a^{\frac{4}{3}}b^{\frac{8}{3}} \tag{5.36}$$

So we have

$$-a^3(a+b) \le \left(-\frac{1}{2}a^{\frac{4}{3}} + b^{\frac{4}{3}}\right)^3 \tag{5.37}$$

Then,

$$-a(a+b)^{\frac{1}{3}} \le -\frac{1}{2}a^{\frac{4}{3}} + b^{\frac{4}{3}} \tag{5.38}$$

This is the proof.

5.2.2 Adaptive NNs Control

The ideal virtual control is

$$\begin{aligned} &\alpha_1^* = -k_1 z_1^\eta - f_1 + \dot{y}_d \\ &\alpha_i^* = -z_{i-1} - k_i z_i^\eta - f_i + \dot{\alpha}_{i-1}^* \\ &2 \le i \le n-1 \\ &\alpha_n^* = u - z_{n-1} - k_n z_n^\eta - f_n + \dot{\alpha}_{n-1}^* \end{aligned} \tag{5.39}$$

In the first step, consider system

$$z_1 = x_1 - y_d \tag{5.40}$$

Then, it has

$$\dot{z}_1 = f_1 + x_2 - \dot{y}_d \tag{5.41}$$

The ideal virtual control law

$$\alpha_1^* = -k_1 z_1^\eta - f_1 + \dot{y}_d \tag{5.42}$$

Choose the NN to approximate the nonlinear system f_1, $\overline{x}_2 \in \Omega_1 \subset \Re^2$ and Ω_1 is compact set

$$f_1(\overline{x}_2(t)) = W_1^{T*} \Psi_1(\overline{x}_2(t)) - \varepsilon_1(t) \tag{5.43}$$

Therefore,

$$\dot{z}_1(t) = W_1^{*T} \Psi_1(\overline{x}_2(t)) - \varepsilon_1(t) \tag{5.44}$$

where $\Gamma_1 = \Gamma_1^T > 0$, and $\sigma_1 > 0$ is positive constant design parameters.

Choose the practical virtual control law

$$\alpha_1 = -k_1 z_1^\eta - \widehat{W}_1^T \Psi_1(\overline{x}_2(t)) \tag{5.45}$$

Choose the adaptive law

$$\dot{\widehat{W}}_1 = \Gamma_1 \left[z_1 \Psi_1(\overline{x}_1(t)) - \sigma_1 \widehat{W}_1^\eta \right] \tag{5.46}$$

Because of

$$x_2 = \alpha_1 + z_2 \tag{5.47}$$

then

$$\begin{aligned} \dot{z}_1 &= \dot{x}_1 \\ &= \left(W_1^{*T} - \widehat{W}_1^T \right) \Psi_1(\overline{x}_1(t)) - k_1 z_1^\eta + z_2 - \varepsilon_1(t) \\ &= -\widetilde{W}_1^T \Psi_1(\overline{x}_1(t)) - k_1 z_1^\eta + z_2 - \varepsilon_1(t) \end{aligned} \tag{5.48}$$

where

$$\widetilde{W}_1 = \widehat{W}_1 - W_1^* \tag{5.49}$$

From the following inequality,

$$-\sigma_1 \widetilde{W}_1^{\mathrm{T}} \widehat{W}_1^{\eta} \le -\frac{\sigma_1}{2} \widetilde{W}_1^{2\eta \mathrm{T}} \widetilde{W}_1^{2\eta} + \sigma_1 W_1^{*2\eta \mathrm{T}} W_1^{*2\eta} \tag{5.50}$$

Choose the Lyapunov candidate function

$$V_1 = \frac{1}{2} z_1^2 + \frac{1}{2} \widetilde{W}_1^{\mathrm{T}} \Gamma_1^{-1} \widetilde{W}_1 \tag{5.51}$$

Then,

$$\begin{aligned} \dot{V}_1 &= z_1 \dot{z}_1 + \widetilde{W}_1^{\mathrm{T}} \Gamma_1^{-1} \dot{\widetilde{W}}_1 \\ &= -k_1 z_1^{4\eta} + z_1 z_2 - z_1 \varepsilon_1(t) - \sigma_1 \widetilde{W}_1^{\mathrm{T}} \widehat{W}_1^{\eta} \end{aligned} \tag{5.52}$$

$$z_1 \varepsilon_1 \le |z_1||\varepsilon_1| \le \frac{3}{4} z_1^{4\eta} + \frac{1}{4} \varepsilon_1^4 \tag{5.53}$$

$$-\sigma_1 \widetilde{W}_1^{\mathrm{T}} \widehat{W}_1^{\eta} \le -\frac{\sigma_1}{2} \widetilde{W}_1^{2\eta \mathrm{T}} \widetilde{W}_1^{2\eta} + \sigma_1 W_1^{*2\eta \mathrm{T}} W_1^{*2\eta} \tag{5.54}$$

$$\dot{V}_1 = -\left(k_1 - \frac{3}{4}\right) z_1^{4\eta} + z_1 z_2 - \frac{\sigma_1}{2} \widetilde{W}_1^{2\eta \mathrm{T}} \widetilde{W}_1^{2\eta} + \frac{1}{4} \varepsilon_1^4 + \sigma_1 W_1^{*2\eta \mathrm{T}} W_1^{*2\eta} \tag{5.55}$$

$$\begin{aligned} \dot{V}_1 &\le z_1 z_2 - \eta_1 \left(\frac{1}{2} z_1^2\right)^{2\eta} - \eta_1 \left(\frac{1}{2} \widetilde{W}_1^{\mathrm{T}} \Gamma_1^{-1} \widetilde{W}_1\right)^{2\eta} + \frac{1}{4} \varepsilon_1^4 \\ &\quad + \sigma_1 W_1^{*2\eta \mathrm{T}} W_1^{*2\eta} \\ &\le z_1 z_2 - \eta_1 V_1^{2\eta} + \delta_1 \end{aligned} \tag{5.56}$$

$$\begin{aligned} \eta_1 &= \min\left(4^{\eta}\left(k_1 - \frac{3}{4}\right), \frac{\sigma_1}{2^{\eta}(\lambda_{\max}(\Gamma_1))^{2\eta}}\right) \\ \delta_1 &= \frac{1}{4} \varepsilon_1^4 + \sigma_1 W_1^{*2\eta \mathrm{T}} W_1^{*2\eta} \end{aligned} \tag{5.57}$$

The ith step $2 \le i \le n-1$

$$z_i = x_i - \alpha_{i-1} \tag{5.58}$$

Choose the virtual control law

$$\alpha_i^* = -z_{i-1} - k_i z_i^{\eta} - f_i + \dot{\alpha}_{i-1} \tag{5.59}$$

Choose the NN to approximate the nonlinear system $f_i(\overline{x}_i(t)) - \dot{\alpha}_{i-1}$, $\overline{x}_i \in \Omega_i \subset \Re^i$, and Ω_i is compact set

$$f_i(\overline{x}_i(t)) - \dot{\alpha}_{i-1} = W_i^{*T} \Psi_i(\overline{x}_i(t)) - \varepsilon_i(t) \tag{5.60}$$

where

$$\dot{\alpha}_{i-1} = \sum_{j=1}^{i} \frac{\partial \alpha_{i-1}}{\partial x_j} \dot{x}_j + \sum_{j=1}^{i-1} \frac{\partial \alpha_{i-1}}{\partial \widehat{W}_j^T} \dot{\widehat{W}}_j^T \tag{5.61}$$

Based on the system

$$\dot{z}_i = W_i^{*T} \Psi_i(\overline{x}_{i+1}(t)) - \varepsilon_i(t) \tag{5.62}$$

Choose the adaptive law

$$\dot{\widehat{W}}_i = \Gamma_i \left[z_i \Psi_i(\overline{x}_i(t)) - \sigma_i \widehat{W}_i^{\eta} \right] \tag{5.63}$$

where $\Gamma_i = \Gamma_i^T > 0$, and $\sigma_i > 0$ is positive constant design parameters.

Choose the practical virtual control law

$$\alpha_i = -z_{i-1} - k_i z_i^{\eta} - \widehat{W}_i^T \Psi_i(\overline{x}_i(t)) \tag{5.64}$$

Then,

$$\begin{aligned} \dot{z}_i &= \left(W_i^* - \widehat{W}_i\right) \Psi_i(\overline{x}_i(t)) - z_{i-1} - k_i z_i^{\eta} + z_{i+1} - \varepsilon_i(t) \\ &= -\widetilde{W}_i \Psi_i(\overline{x}_i(t)) - z_{i-1} - k_i z_i^{\eta} + z_{i+1} - \varepsilon_i(t) \end{aligned} \tag{5.65}$$

where

$$\widetilde{W}_i = \widehat{W}_i - W_i^* \tag{5.66}$$

Choose the Lyapunov candidate function

$$V_i = \frac{1}{2} z_i^2 + \frac{1}{2} \widetilde{W}_i^{\mathrm{T}} \Gamma_i^{-1} \widetilde{W}_i \tag{5.67}$$

Then,

$$\begin{aligned} \dot{V}_i &= z_i \dot{z}_i + \widetilde{W}_i \Gamma_i^{-1} \dot{\widetilde{W}}_i^{\mathrm{T}} \\ &= -z_{i-1} z_i - k_i z_i^{4\eta} + z_i z_{i+1} - z_i \varepsilon_i(t) - \sigma_i \widetilde{W}_i^{\mathrm{T}} \widehat{W}_i^{\eta} \end{aligned} \tag{5.68}$$

From the following inequality

$$z_i\varepsilon_i \le |z_i||\varepsilon_i| \le \frac{3}{4}z_i^{4\eta} + \frac{1}{4}\varepsilon_i^4 \tag{5.69}$$

$$-\sigma_i \widetilde{W}_i^{\mathrm{T}} \widehat{W}_i^{2\eta} \le -\frac{\sigma_i}{2}\widetilde{W}_i^{2\eta\mathrm{T}} \widetilde{W}_i^{2\eta} + \sigma_i W_i^{*2\eta\mathrm{T}} W_i^{*2\eta} \tag{5.70}$$

$$\begin{aligned} \dot{V}_i &= z_i\dot{z}_i + \widetilde{W}_i^{\mathrm{T}} \Gamma_i^{-1} \dot{\widetilde{W}}_i \\ &= -z_{i-1}z_i - \left(k_i - \frac{3}{4}\right) z_i^{4\eta} + z_i z_{i+1} - \frac{\sigma_i}{2}\widetilde{W}_i^{2\eta\mathrm{T}} \widetilde{W}_i^{2\eta} + \frac{1}{4}\varepsilon_i^4 \\ &\quad + \sigma_i W_i^{*2\eta\mathrm{T}} W_i^{*2\eta} \end{aligned} \tag{5.71}$$

$$\begin{aligned} \dot{V}_i &\le -z_{i-1}z_i + z_i z_{i+1} - \eta_i \left(\frac{1}{2}z_i^2\right)^{2\eta} - \eta_i \left(\frac{1}{2}\widetilde{W}_i^{\mathrm{T}} \Gamma_i^{-1} \widetilde{W}_i\right)^{2\eta} + \frac{1}{4}\varepsilon_i^4 \\ &\quad + \sigma_i W_i^{*2\eta\mathrm{T}} W_i^{*2\eta} \\ &\le -z_{i-1}z_i + z_i z_{i+1} - \eta_i V_i^{2\eta} + \delta_i \end{aligned} \tag{5.72}$$

$$\begin{aligned} \eta_i &= \min\left(4^{\eta}\left(k_i - \frac{3}{4}\right), \frac{\sigma_i}{2^{\eta}(\lambda_{\max}(\Gamma_i))^{2\eta}}\right) \\ \delta_i &= \frac{1}{4}\varepsilon_i^4 + \sigma_i W_i^{*2\eta\mathrm{T}} W_i^{*2\eta} \end{aligned} \tag{5.73}$$

The nth step, this is the most important step. Based on the system

$$z_n = x_n - \alpha_{n-1} \tag{5.74}$$

Choose the NN to approximate the nonlinear system $f_n(\overline{x}_n(t), u) - \dot{\alpha}_{n-1}$, $\overline{x}_n^T \in \Omega_n \subset \Re^n$, and Ω_n is compact set

$$f_n(\overline{x}_n(t), u) - \dot{\alpha}_{n-1} = W_n^{*T} \Psi_n(\overline{x}_n(t), u) - \varepsilon_n(t) \tag{5.75}$$

where

$$\dot{\alpha}_{n-1} = \sum_{j=1}^{n} \frac{\partial \alpha_{n-1}}{\partial x_j}\dot{x}_j + \sum_{j=1}^{n-1} \frac{\partial \alpha_{n-1}}{\partial \widehat{W}_j^T}\dot{\widehat{W}}_j^T \tag{5.76}$$

Based on the system

$$\dot{z}_n = W_n^{*T}\Psi_n(\overline{x}_n(t), u) - \varepsilon_n(t) \tag{5.77}$$

Choose the adaptive law

$$\dot{\widehat{W}}_n = \Gamma_n\left[z_n\Psi_n^T(\overline{x}_n) - \sigma_n\widehat{W}_n^{\eta}\right] \tag{5.78}$$

where $\Gamma_n = \Gamma_n^T > 0$, and $\sigma_n > 0$ is positive constant design parameters.

$$\alpha_n = -z_{n-1} - k_n z_n^{\eta} + u - \widehat{W}_n^T\Psi_n(\overline{x}_{n+1}(t)) \tag{5.79}$$

Then,

$$\dot{z}_n = -\widetilde{W}_n^T\Psi_n(\overline{x}_n(t)) - z_{n-1} - k_n z_n^{\frac{1}{3}} + z_{n+1} - \varepsilon_n(t) \tag{5.80}$$

where

$$\widetilde{W}_n = \widehat{W}_n - W_n^* \tag{5.81}$$

From the following inequality,

$$-\sigma_n\widetilde{W}_n^T\widehat{W}_n^{\frac{1}{3}} \leq -\frac{\sigma_n}{2}\widetilde{W}_n^{\frac{2}{3}T}\widetilde{W}_n^{\frac{2}{3}} + \sigma_n W_n^{*\frac{2}{3}T}W_n^{*\frac{2}{3}} \tag{5.82}$$

Choose the Lyapunov candidate function

$$V_n = \frac{1}{2}z_n^2 + \frac{1}{2}\widetilde{W}_n\Gamma_n^{-1}\widetilde{W}_n^{\mathrm{T}} \tag{5.83}$$

Then,

$$\begin{aligned}\dot{V}_n &= z_n\dot{z}_n + \widetilde{W}_n\Gamma_n^{-1}\dot{\widetilde{W}}_n^{\mathrm{T}} \\ &= -z_{n-1}z_n - k_n z_n^{\frac{4}{3}} + z_n z_{n+1} - z_n\varepsilon_n(t) - \sigma_n\widetilde{W}_n^{\mathrm{T}}\widehat{W}_n^{\frac{1}{3}}\end{aligned} \tag{5.84}$$

$$z_n\varepsilon_n \leq |z_n||\varepsilon_n| \leq \frac{3}{4}z_n^{\frac{4}{3}} + \frac{1}{4}\varepsilon_n^4 \tag{5.85}$$

$$-\sigma_n\widetilde{W}_n^T\widehat{W}_n^{\frac{1}{3}} \leq -\frac{\sigma_n}{2}\widetilde{W}_n^{\frac{2}{3}T}\widetilde{W}_n^{\frac{2}{3}} + \sigma_n W_n^{*\frac{2}{3}T}W_n^{*\frac{2}{3}} \tag{5.86}$$

$$\dot{V}_n \leq -z_{n-1}z_n - \left(k_n - \frac{3}{4}\right)z_n^{\frac{4}{3}} + z_n z_{n+1} - \frac{\sigma_n}{2}\widetilde{W}_n^{\frac{2}{3}\mathrm{T}}\widetilde{W}_n^{\frac{2}{3}} + \frac{1}{4}\varepsilon_n^4$$
$$+ \sigma_n W_n^{*\frac{2}{3}\mathrm{T}} W_n^{*\frac{2}{3}} \tag{5.87}$$

$$\dot{V}_n \leq -z_{n-1}z_n + z_n z_{n+1} - \eta_n\left(\frac{1}{2}z_n^2\right)^{\frac{2}{3}} - \eta_n\left(\frac{1}{2}\widetilde{W}_n^{\mathrm{T}}\Gamma_n^{-1}\widetilde{W}_n\right)^{\frac{2}{3}} + \delta_n$$
$$\leq -z_{n-1}z_n + z_n z_{n+1} - \eta_n V_n^{\frac{2}{3}} + \delta_n \tag{5.88}$$

where

$$\eta_n = \min\left(4^{\frac{1}{3}}\left(k_n - \frac{3}{4}\right), \frac{\sigma_n}{2^{\frac{1}{3}}(\lambda_{\max}(\Gamma_n))^{\frac{2}{3}}}\right)$$
$$\delta_n = \frac{1}{4}\varepsilon_n^4 + \sigma_n W_n^{*\frac{2}{3}T} W_n^{*\frac{2}{3}} \tag{5.89}$$

The $n+1$th step, this is the most important step. Based on the system

$$z_{n+1} = u - \alpha_n \tag{5.90}$$

$$\dot{z}_{n+1} = v - \dot{\alpha}_n \tag{5.91}$$

Choose the NN to approximate the nonlinear system $-\dot{\alpha}_n(\overline{x}_{n+1})$, $\overline{x}_{n+1}^T = \begin{bmatrix}\overline{x}_n^T & u\end{bmatrix} \in \Omega_{n+1} \subset \Re^{n+1}$ and Ω_{n+1} is compact set

$$-\dot{\alpha}_n = W_{n+1}^{*T}\Psi_{n+1}(\overline{x}_{n+1}(t)) - \varepsilon_{n+1}(t) \tag{5.92}$$

where

$$\dot{\alpha}_n = \sum_{j=1}^{n}\frac{\partial \alpha_n}{\partial x_j}\dot{x}_j + \sum_{j=1}^{n}\frac{\partial \alpha_n}{\partial \widehat{W}_j^T}\dot{\widehat{W}}_j^T \tag{5.93}$$

Based on the system

$$\dot{z}_{n+1} = v + W_{n+1}^{*T}\Psi_{n+1}(\overline{x}_{n+1}(t)) - \varepsilon_{n+1}(t) \tag{5.94}$$

Choose the adaptive law

$$\dot{\widehat{W}}_{n+1} = \Gamma_{n+1}\left[z_{n+1}\Psi_{n+1}(\overline{x}_{n+1}) - \sigma_{n+1}\widehat{W}^{\frac{1}{3}}_{n+1}\right] \tag{5.95}$$

where $\Gamma_{n+1} = \Gamma^T_{n+1} > 0$, and $\sigma_{n+1} > 0$ is positive constant design parameters.

$$v = -z_n - k_{n+1}z^{\frac{1}{3}}_{n+1} - \widehat{W}^T_{n+1}\Psi_{n+1}(\overline{x}_{n+1}(t)) \tag{5.96}$$

Then,

$$\dot{z}_{n+1} = -z_n - k_{n+1}z^{\frac{1}{3}}_{n+1} - \widetilde{W}^T_{n+1}\Psi_{n+1}(\overline{x}_{n+1}(t)) - \varepsilon_{n+1}(t) \tag{5.97}$$

where

$$\widetilde{W}_{n+1} = \widehat{W}_{n+1} - W^*_{n+1} \tag{5.98}$$

From the following inequality

$$-\sigma_{n+1}\widetilde{W}^T_{n+1}\widehat{W}^{\frac{1}{3}}_{n+1} \le -\frac{\sigma_{n+1}}{2}\widetilde{W}^{\frac{2}{3}T}_{n+1}\widetilde{W}^{\frac{2}{3}}_{n+1} + \sigma_{n+1}W^{*\frac{2}{3}T}_{n+1}W^{*\frac{2}{3}}_{n+1} \tag{5.99}$$

Choose the Lyapunov candidate function

$$V_{n+1} = \frac{1}{2}z^2_{n+1} + \frac{1}{2}\widetilde{W}_{n+1}\Gamma^{-1}_{n+1}\widetilde{W}^{\mathrm{T}}_{n+1} \tag{5.100}$$

then

$$\begin{aligned} \dot{V}_{n+1} &= z_{n+1}\dot{z}_{n+1} + \widetilde{W}_{n+1}\Gamma^{-1}_{n+1}\dot{\widetilde{W}}^{\mathrm{T}}_{n+1} \\ &= -z_n z_{n+1} - k_{n+1}z^{\frac{4}{3}}_{n+1} - z_{n+1}\varepsilon_{n+1}(t) - \sigma_{n+1}\widetilde{W}^{\mathrm{T}}_{n+1}\widehat{W}^{\frac{1}{3}}_{n+1} \end{aligned} \tag{5.101}$$

where

$$z_{n+1}\varepsilon_{n+1} \le |z_{n+1}||\varepsilon_{n+1}| \le \frac{3}{4}z^{\frac{4}{3}}_{n+1} + \frac{1}{4}\varepsilon^4_{n+1} \tag{5.102}$$

$$-\sigma_{n+1}\widetilde{W}^{\mathrm{T}}_{n+1}\widehat{W}^{\frac{1}{3}}_{n+1} \le -\frac{\sigma_{n+1}}{2}\widetilde{W}^{\frac{2}{3}T}_{n+1}\widetilde{W}^{\frac{2}{3}}_{n+1} + \sigma_{n+1}W^{*\frac{2}{3}T}_{n+1}W^{*\frac{2}{3}}_{n+1} \tag{5.103}$$

$$\begin{aligned} \dot{V}_{n+1} \le {} & -z_n z_{n+1} - \left(k_{n+1} - \frac{3}{4}\right)z^{\frac{4}{3}}_{n+1} - \frac{\sigma_{n+1}}{2}\widetilde{W}^{\frac{2}{3T}}_{n+1}\widetilde{W}^{\frac{2}{3}}_{n+1} + \frac{1}{4}\varepsilon^4_{n+1}(t) \\ & + \sigma_{n+1}W^{*\frac{2}{3T}}_{n+1}W^{*\frac{2}{3}}_{n+1} \end{aligned} \tag{5.104}$$

$$\begin{aligned}\dot{V}_{n+1} &\leq -z_n z_{n+1} - \eta_{n+1}\left(\frac{1}{2}z_{n+1}^2\right)^{\frac{2}{3}} - \eta_{n+1}\left(\frac{1}{2}\widetilde{W}_{n+1}^{\mathrm{T}}\Gamma_{n+1}^{-1}\widetilde{W}_{n+1}\right)^{\frac{2}{3}} \\ &\quad + \delta_{n+1} \\ &\leq -z_n z_{n+1} - \eta_{n+1}V_{n+1}^{\frac{2}{3}} + \delta_{n+1}\end{aligned} \tag{5.105}$$

$$\begin{aligned}\eta_{n+1} &= \min\left(4^{\frac{1}{3}}\left(k_{n+1} - \frac{3}{4}\right), \frac{\sigma_{n+1}}{2^{\frac{1}{3}}(\lambda_{\max}(\Gamma_{n+1}))^{\frac{2}{3}}}\right) \\ \delta_{n+1} &= \frac{1}{4}\varepsilon_{n+1}^4 + \sigma_{n+1}W_{n+1}^{*\frac{2}{3}\mathrm{T}}W_{n+1}^{*\frac{2}{3}}\end{aligned} \tag{5.106}$$

$$V = \sum_{i=1}^{n+1} V_i \tag{5.107}$$

Then, based on this, we have

$$\dot{V} \leq -\eta V^{\frac{2}{3}} + \delta \tag{5.108}$$

where

$$\begin{aligned}\eta &= \min(\eta_i), i = 1, 2, 3 \cdots n+1 \\ \delta &= \sum_{i=1}^{n+1} \delta_i\end{aligned} \tag{5.109}$$

Theorem 5.2: Under homeomorphism mapping and the above assumptions, consider the closed-loop adaptive system consisting of plant and control law with the virtual control and adaptive laws. Then, for any bounded initial conditions, the following statements hold:

1. All the signals of the closed loop system, including x_i, α_i, $\widehat{W}_i$, remain bounded all the time.
2. The closed loop signal $z = \begin{bmatrix} z_1 & z_2 & \cdots & z_{n+1}\end{bmatrix}^T \in \Re^{n+1}$ converge to a compact set defined by

$$\Omega_s = \left\{ z \middle| \|z\|^2 \leq \frac{\delta}{\eta} \right\} \tag{5.110}$$

In finite time, where δ,η are constants related to the design parameters.

3. The finite time T is given by

$$T \leq \frac{3V^{\frac{1}{3}}(x_0)}{\eta\theta_0} \tag{5.111}$$

where η, θ_0 are constants related to the design parameters, $V(x_0)$ is constant related to Lyapunov candidate functional.

Proof: Considering Lyapunov function candidate, then V is bounded in finite time. Therefore, it can conclude that for all $1 \leq i \leq n+1$, the error signals $z_i\widetilde{\mathrm{Wi}}$ are UUB in finite time.

5.2.3 Simulation Example

In this section, two examples are considered to show the generalization of the proposed adaptive finite-time neural networks control.

5.2.3.1 Mathematical Example

In this section, a mathematical example is presented to illustrate the validity of the proposed adaptive finite-time NN control approach.

$$\begin{aligned}
\dot{x}_1 &= x_1 e^{-x_1^2} + x_2^3 + x_2 e^{-1-x_1^2} \\
\dot{x}_2 &= x_1^2 + 0.1\left(1 + x_2^2\right)u + 0.15u^3 + \sin(u) \\
y &= x_1
\end{aligned} \tag{5.112}$$

where $x = \left[x_1 \quad x_2\right]^T$, u and y are state vector, control input and output, respectively; the reference output is taken as $y_d = 5\sin(t)$. The control objective is to design the adaptive neural tracking control for system such that all the signals in the closed-loop system remain bounded, and the system output follows the given reference signal y_d in finite time.

Based on Theorem 5.2, the adaptive neural networks control law is chosen as

$$\dot{u} = -z_2 - k_3 z_3^{\frac{1}{3}} - \widehat{W}_3^T \Psi_3(\overline{x}_2, u) \tag{5.113}$$

The virtual control variables are designed as

$$\alpha_1 = -k_1 z_1^{\frac{1}{3}} + x_2 - \widehat{W}_1^T \Psi_1(\overline{x}_2(t)) \tag{5.114}$$

and

$$\alpha_2 = -z_1 - k_2 z_2^{\frac{1}{3}} + u - \widehat{W}_2^T \Psi_2(\overline{x}_2, u) \tag{5.115}$$

The adaptive laws are given as follows, for $i = 1, 2, 3$

$$\dot{\widehat{W}}_i = \Gamma_i \left[z_i \Psi_i(\overline{x}_{i+1}) - \sigma_i \widehat{W}_i^{\frac{1}{3}} \right] \tag{5.116}$$

with $z_1 = x_1 - y_d$, $z_2 = x_2 - \alpha_1$, and $z_3 = u - \alpha_2$.

In the simulation studies, Gaussian RBF neural networks $\widehat{W}^T \Psi(x)$ were selected. The structure of the first RBF is 1−7−1, with the center μ_i evenly placed on $[-3 \quad 3]$, such as $[-3 \quad -2 \quad -1 \quad 0 \quad 1 \quad 2 \quad 3]$, and the width $\eta = 0.46$. The structure of the second RBF is 2−9−1, with the center μ_i evenly placed on $[-9 \quad 9]$, such as

$$\begin{bmatrix} -9 & -7 & -5 & -3 & 0 & 3 & 5 & 7 & 9 \\ -9 & -7 & -5 & -3 & 0 & 3 & 5 & 7 & 9 \end{bmatrix} \tag{5.117}$$

And the width is $\eta = 10$. The structure of the third RBF is 2−9−1, with the center μ_i evenly placed on $[-9 \quad 9]$, such as

$$\begin{bmatrix} -9 & -7 & -5 & -3 & 0 & 3 & 5 & 7 & 9 \\ -9 & -7 & -5 & -3 & 0 & 3 & 5 & 7 & 9 \end{bmatrix} \tag{5.118}$$

And the width is $\eta = 10$. The design parameters are taken as $\Gamma_1 = diag\{0.5\}; \Gamma_2 = diag\{10\}$ $\Gamma_3 = diag\{5\}$ all initial conditions are set as zero.

The simulation results are shown in Figs. 5.10-5.13. It can be observed from Fig. 5.10 that the tracking error satisfies the requirement exactly after 3 s. Fig. 5.11 shows the trajectory of control input, bounded. Figs. 5.12 and 5.13 show the trajectories of f_1, NN_1 and $f_2 - \dot{\alpha}_1, NN_2$, respectively, and the tracking errors satisfy the requirement.

From Fig. 5.12, the tracking error and chattering only exists in the beginning. This is due to the lack of knowledge about the plant nonlinearities and the initial weights chosen randomly. Through the NN learning phase, better tracking performance is obtained, and the chattering also disappears after 5 s.

To illustrate the learning ability of NNs, the nonlinear function f_1 and its estimate NN_1 is shown in Fig. 5.12, and the nonlinear function $f_2 - \dot{\alpha}_1$ and its estimate NN_2 is shown in Fig. 5.13.

5.2.3.2 Plate-Ball Example

Consider a Plate-Ball model (Yuan and Zhang, 2010) in form of

$$\begin{cases} \left(m + \frac{I_b}{r_b^2}\right)\ddot{\xi}_x + mgu_x + f_x = 0 \\ \left(m + \frac{I_b}{r_b^2}\right)\ddot{\xi}_y + mgu_y + f_y = 0 \end{cases} \tag{5.119}$$

where m is ball mass, I_b is ball moment of inertia, r_b is ball radius, u_x, u_y are plate inclination angle in x or y direction, and f_x, f_y contain the centrifugal torque resulting from rotation and other modeling error, which can be omitted.

Choose the homeomorphism

$$\begin{aligned} \xi_x &= \frac{2a}{\pi}\arctan(x) \\ \xi_y &= \frac{2b}{\pi}\arctan(y) \end{aligned} \tag{5.120}$$

Then,

$$\begin{aligned} x &= tan\left(\frac{\pi\xi_x}{2a}\right) \\ y &= tan\left(\frac{\pi\xi_y}{2b}\right) \end{aligned} \tag{5.121}$$

Therefore,

$$\ddot{x} = \frac{\pi^2}{2a^2}\tan\left(\frac{\pi\xi_x}{2a}\right)\left(1+\tan^2\left(\frac{\pi\xi_x}{2a}\right)\right)\ddot{\xi}_x \tag{5.122}$$

Then,

$$\ddot{\xi}_x = \frac{2a^2}{\pi^2 x(1+x^2)}\ddot{x} \tag{5.123}$$

Then, system (Eq. 5.119) transform into

$$\left(m+\frac{I_b}{r_b^2}\right)\frac{2a^2}{\pi^2 x(1+x^2)}\ddot{x} + mgu_x + f_x = 0 \tag{5.124}$$

Let $x_1 = x, x_2 = \dot{x}$, then the above equations can be changed as

$$\begin{cases} \dot{x}_1 = x_2 \\ \dot{x}_2 = f_x - \dfrac{mgr_b^2\pi^2 x_1\left(1+x_1^2\right)}{2a^2\left(mr_b^2+I_b\right)}u_x \end{cases} \tag{5.125}$$

Again, it was used to verify the adaptation capability of the proposed, adaptive, finite-time, neural network control, choose the $f_x = -x_1 - 2x_2$ as modeling error in simulation process.

In simulation, to illustrate the adaptation capability of the proposed adaptive NNs control, RBF NNs are used with the same centers, width, and the number of NN nodes with Example A. The simulation results are shown in Fig. 5.14, and the output of the designed system satisfies the required accuracy. Obviously, simulation results show that the controller works well and achieves the desired convergence performance. From homeomorphism mapping skill, the finite states of Plate-Ball map to infinite states, then the control design by infinite system can be used to control the Plate-Ball system state constraint and convergence in finite time, respectively.

REFERENCES

Ge, S.S., Hong, F., Tong, H.L., 2003. Adaptive neural network control of nonlinear systems with unknown time delays. IEEE Transactions on Automatic Control 48, 2004–2010.

Ge, S.S., Hong, F., Lee, T.H., 2005. Robust adaptive control of nonlinear systems with unknown time delays. Automatica 41, 1181–1190.

Jes, D.S., Rubio, J., Yu, W., 2007. Nonlinear system identification with recurrent neural networks and dead-zone Kalman filter algorithm. Neurocomputing 70, 2460–2466.

Lewis, F.L., Yegildirek, A., Liu, K., 1996. Multilayer Neural-net Robot Controller with Guaranteed Tracking Performance. IEEE Press.

Li, Y., Zhang, J., Wu, X., Zhu, Q., 2016. Adaptive neural network sliding mode control for a class of non-affine nonlinear systems with pure-feedback prototype. In: International Conference on Modelling, Identification and Control, pp. 1–6.

Liu, S., Wang, Y., Zhu, Q., 2008. Development of a new EDRNN procedure in control of human arm trajectories. Neurocomputing 72, 490–499.

Polycarpou, M.M., 1996. Stable adaptive neural scheme for nonlinear systems. IEEE Transactions on Automatic Control 41 (3), 447–451.

Rubio, J.D.J., Yu, W., 2007. Stability analysis of nonlinear system identification via delayed neural networks. IEEE Transactions on Circuits and Systems II Express Briefs 54, 161–165.

Rui Peng, L.C., Yating, D., Xitong, Z., Huaiyu, W., Yang, C., 2018. Design and implementation of monocular vision SLAM for mobile robot based on ROS. International Journal of Applied Mathematics in Control Engineering 1, 85–91.

Shengli Shi, M.L., Jiayu, M.A., 2018. Fractional-order sliding mode control for a class of nonlinear systems via nonlinear disturbance observer. International Journal of Applied Mathematics in Control Engineering 1.

Wang, D., Huang, J., 2002. Brief Adaptive Neural Network Control for a Class of Uncertain Nonlinear Systems in Pure-feedback Form. Pergamon Press, Inc.

Wu, X., Zhang, J., Zhu, Q., 2010. A generalized procedure in designing recurrent neural network identification and control of time-varying-delayed nonlinear dynamic systems. Neurocomputing 73, 1376–1383.

Yafei Liu, J.N., Jun, Y., Guanbin, G.A.O., 2018. Modified CRM-based model reference adaptive control with reduced peaking phenomenon. International Journal of Applied Mathematics in Control Engineering 1, 70–78.

Yu, W., 2004. Nonlinear system identification using discrete-time recurrent neural networks with stable learning algorithms. Information Sciences 158, 131–147.

Yu, W., 2006. Multiple recurrent neural networks for stable adaptive control. Neurocomputing 70, 430–444.

Yuan, D., Zhang, Z., 2010. Modelling and control scheme of the ball-plate trajectory-tracking pneumatic system with a touch screen and a rotary cylinder. IET Control Theory and Applications 4, 573–589.

Zhang, T., Ge, S.S., Hang, C.C., 1999. Design and performance analysis of a direct adaptive controller for nonlinear systems ☆. Automatica 35, 1809–1817.

Zhang, L.F., Zhu, Q.M., Longden, A., 2009. A correlation-test-based validation procedure for identified neural networks. IEEE Transactions on Neural Networks 20, 1–13.

Zhang, J., Zhu, Q., Wu, X., Li, Y., 2013. Letters: a generalized indirect adaptive neural networks backstepping control procedure for a class of non-affine nonlinear systems with pure-feedback prototype. Neurocomputing 121, 131–139.

CHAPTER 6
Applications

Contents

Chapter 6, first, studies the intelligent flight task algorithm for aircraft to effectively determine and search the best flight routes and complicated condition, where the condition includes the complicated topography and threaten from other flights in the route airspace. Second, it studies the multilateration system theory and positioning algorithm using back propagation (BP) network (Fig. 6.1).

6.1 AIRCRAFT PATH PLANNING BASED ON NEURAL NETWORKS

Aircraft path planning is a required process when guiding an aircraft to move around obstacles; these processes include static and dynamic obstacles. Many different methods have been introduced for path planning, such as mathematical programming methods and evolutionary algorithms (Cimino and Mirabelli, 2009; Besada-Portas et al., 2013; Qiang Chen and

Adaptive Sliding Mode Neural Network Control for Nonlinear Systems
ISBN 978-0-12-815372-7
https://doi.org/10.1016/B978-0-12-815372-7.00006-9

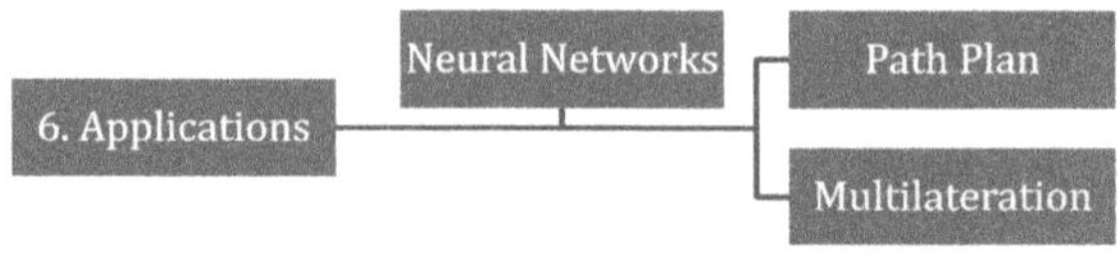

Figure 6.1 The structure chart of Chapter 6.

Zhongjun, 2018). The potential field method, on behalf of mathematical programming method, which create a scalar field with artificial obstacles and forced to avoid obstacles toward the destination, may make the object trapped in a local area. It is hard to solve complex optimization problems characterized by multimodality, discontinuity, and high dimensionality (Wang et al., 2015; Sun et al., 2011).

The powerful evolutionary programming techniques have considerable potential to be investigated in the pursuit for more efficient algorithms because the path-planning problem is basically an optimal search problem (Espelosín et al., 2013). Evolutionary algorithms (Gosciniak, 2015; Mohemmed et al., 2008; Wu et al., 2014), such as genetic algorithm, ant colony optimization, and particle swarm optimization (PSO), play an important role in path planning. PSO is a competitive population-based algorithm, which is particularly efficient in dealing with numerical optimization problems. However, the evolutionary algorithms require greater number of iterations in the optimization process but are less likely to stop the optimization process in the local optimum. To solve this problem, the powerful artificial neural networks (ANN) (Xie et al., 2014) have been examined to solve the path-planning problem relying on their parallel architecture to provide a fast solution.

To solve the problem of flight safety and improve the accuracy of the collision risk, the traditional EVENT model is improved, and the EVENT model is built using elliptic cylinder as the collision template. Due to the high communication and monitoring performance, broadcast automatic monitoring (ADS-B) (Ali et al., 2015; Mccallie et al., 2011; Zhang and

Yanbo, 2011) has become an important part of the current navigation system.

The path-planning algorithm is proposed based on ANN architecture and wave propagation theorem. The work is organized as follows. In Section 6.1.2, path model based on risk assessment model is briefly discussed, which provided foundation of the path planning. The network architecture is presented in Section 6.1.3 followed by the path-planning algorithm for solving the path-planning problem being provided in Section 6.1.4. The results from computer simulation experiments are discussed in Section 6.1.5.

6.1.1 Path Model Based on Risk Assessment Model

In 2003, Brooker first proposed a model based on the event of the post-Reich model that is EVENT model, and its lateral interval to solve. In 2006, Brooker also studied the longitudinal EVENT model. As shown in Fig. 6.2:

When considering the longitudinal collision risk, the EVENT collision model defines a rectangular collision box for the aircraft A, and B is considered as a particle. Twice the length, width, and height of aircraft are defined as length, width, and height of a cuboid collision box, respectively. With B as the origin of the three-dimensional coordinate system. Two aircraft longitudinal, lateral, and vertical directions are

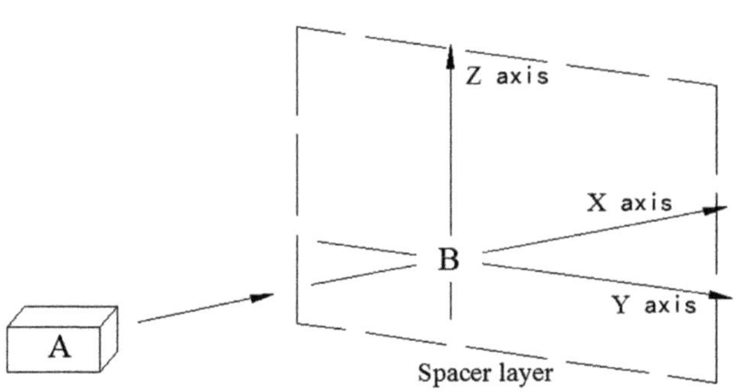

Figure 6.2 Longitudinal EVENT collision risk model.

defined for the *X*, *Y*, and *Z* direction; *Y* and *Z* axes to determine the plane-defined compartments.

As shown in Fig. 6.3, the plane is about to enter the vertical spacer position is *A*, just fly out of spacer position *AA*. *E* to *F* is a crash box that moves in the vertical distance, *H* to *I* is the lateral movement distance. In the process of crossing, the motion area of the collision box *A* is projected to *CDFGH* (shadow part) on the longitudinal spacing layer. If the aircraft B happens to be located in the *CDFGH*, two of the absolute collision occurs. The *CDFGH* shape is more complex, relatively conservative; The *CDFGH* shape is more complex and conservative relatively, and shape *CEGI* defines a rectangle to expand the bounding box, then, if the aircraft is located in the *B* rectangular *CEGI*, two aircrafts would crash. It can be concluded that longitudinal collision risk probability is equal to the bounding box of a longitudinal layer through the interlayer, and the probability of airplane *B* happens to be located in the expansion of the product of the probability of collision box.

The probability value of longitudinal collision risk is

$$P_x = 2E(O)L(S)P_y(O)P_z(O)\left(1+\frac{2u_y\lambda_x}{2u_x\lambda_y}\right)\left(1+\frac{2u_z\lambda_x}{2u_x\lambda_z}\right) \tag{6.1}$$

In formula (6.1), P_x is the probability value of longitudinal collision risk, $E(O)$is the longitudinal approach, that is, the ratio of the number of aircraft in the same direction to the flight of

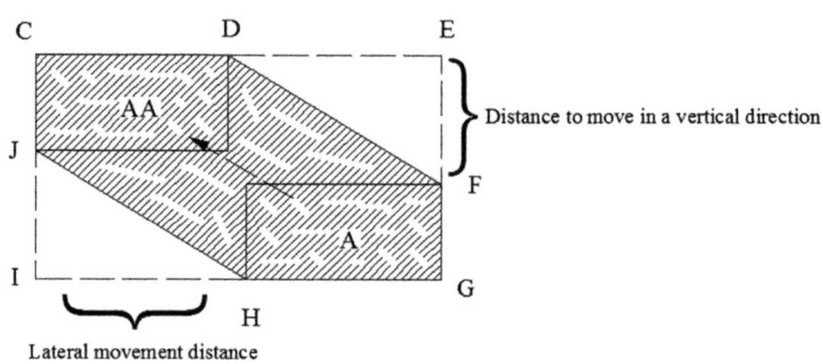

Figure 6.3 Extended collision box.

the adjacent air path is smaller than the number of aircraft in the standard interval and the total number of all aircraft, $L(S)$is the probability of the minimum safety spacing across the aircraft longitudinal spacer, $P_y(O)$ is the probability of lateral overlap of the two aircraft on the same altitude layer, $P_z(O)$is the probability of vertical overlap, λ_x, λ_y, λ_z, respectively, for an aircraft captain, wingspan, and height, u_x, u_y, u_z are vertical, lateral, and vertical direction of the relative velocity.

Event model definition of the bounding box is a cuboid, but in the actual situation, when close to the two planes, aircraft do generally adjust the vertical height to prevent collision, so the aircraft around the short semiaxis deflection formed in the spatial shape closer to an elliptic cylinder. Replacing the traditional rectangular template by the elliptical cylinder collision template is more practical and accurate. Therefore, A as the center of the aircraft, λ_x as the height, λ_y for the long half shaft, and λ_z for the short half axis of the elliptical cylinder. As shown in Fig. 6.4:

Because A travel in vertical spacer layer and the probability of airborne equipment monitoring and navigation performance is related, and the specific shape of the crash box has nothing to do and B is located exactly at the expansion box collision probability, and extended bounding box size is proportional to, therefore, only calculated before and after expansion box collision area proportion of the improvement can be obtained

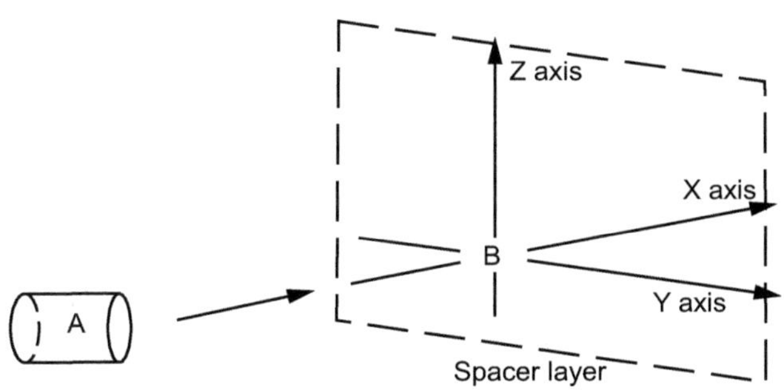

Figure 6.4 Improved EVENT collision risk model.

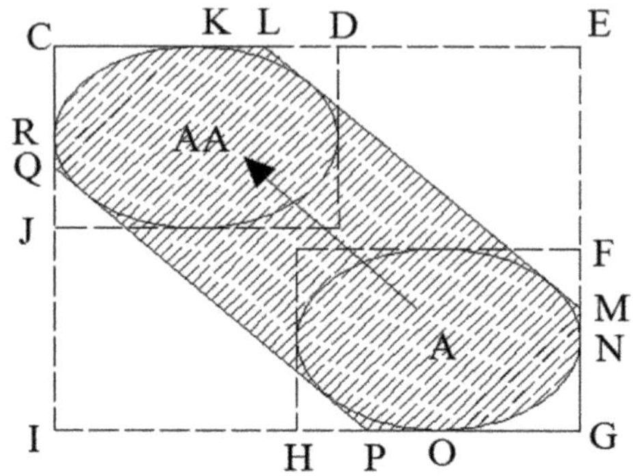

Figure 6.5 Improved collision box.

by improved collision probability. Shown in Fig. 6.5, the shadow region *KLMNOPQR* extended collision box improved ellipsoid.

Located in Fig. 6.3 to expand the area of the collision box for S_1, in Fig. 6.5 to expand the area of the collision box for S_2, and then S_1 and S_2 area ratio of $R(O)$.

EF and *IJ* are the distances in the vertical direction of the collision box *A*:

$$EF = IJ = \frac{2u_z\lambda_x}{u_x} \tag{6.2}$$

DE and *HI* are the distance to the *A* side of the collision case:

$$DE = HI = \frac{2u_y\lambda_x}{u_x} \tag{6.3}$$

In Fig. 6.3,

$$\begin{aligned} CD &= 2\lambda_y \\ FG &= 2\lambda_z \end{aligned} \tag{6.4}$$

So,

$$\begin{aligned} S_1 &= CE \times EG \\ &= (CD + DE)(EF + FG) \\ &= \left(2\lambda_y + \frac{2u_y\lambda_x}{u_x}\right)\left(2\lambda_z + \frac{2u_z\lambda_x}{u_x}\right) \end{aligned} \tag{6.5}$$

Therefore,

$$S_2 = S_1 - S_{CRK} - S_{GON} - S_{\Delta LEM} - S_{\Delta QPI} \tag{6.6}$$

$$\begin{aligned} S_{CRK} = S_{GON} &= CR \times CK - \frac{1}{4}\pi\lambda_y\lambda_z \\ &= \lambda_y\lambda_z - \frac{1}{4}\pi\lambda_y\lambda_z \end{aligned} \tag{6.7}$$

Because the length of the line segment DL is not easy to get, the approximate value method is adopted, $DL = PH = \frac{1}{4}CD = \frac{1}{2}\lambda_y$; similarly, $QJ = FM = \frac{1}{4}FG = \frac{1}{2}\lambda_z$. Therefore,

$$\begin{aligned} S_{\Delta LEM} = S_{\Delta QPI} &= \frac{1}{2}PI \times QI \\ &= \frac{1}{2}\left(\frac{2u_y\lambda_x}{u_x} + \frac{1}{2}\lambda_y\right)\left(\frac{2u_z\lambda_x}{u_x} + \frac{1}{2}\lambda_z\right) \end{aligned} \tag{6.8}$$

Therefore, the formula (6.7) and the formula (6.8) are brought into (6.6), and the area of the elliptical cylinder expansion collision box can be introduced.

$$\begin{aligned} S_2 &= S_1 - S_{CRK} - S_{GON} - S_{\Delta LEM} - S_{\Delta QPI} \\ &= \left(2\lambda_y + \frac{2u_y\lambda_x}{u_x}\right) \times \left(2\lambda_z + \frac{2u_z\lambda_x}{u_x}\right) - 2\lambda_y\lambda_z \\ &+ \frac{1}{2}\pi\lambda_y\lambda_z - \left(\frac{2u_y\lambda_x}{u_x} + \frac{1}{2}\lambda_y\right)\left(\frac{2u_z\lambda_x}{u_x} + \frac{1}{2}\lambda_z\right) \\ &= \frac{3(u_z\lambda_x\lambda_y + u_y\lambda_x\lambda_z)}{u_x} + \left(\frac{1}{2}\pi + \frac{7}{4}\right)\lambda_y\lambda_z \end{aligned} \tag{6.9}$$

So,

$$R(O) = \frac{S_2}{S_1} = \frac{\frac{3(u_z\lambda_x\lambda_y + u_y\lambda_x\lambda_z)}{u_x} + \left(\frac{1}{2}\pi + \frac{7}{4}\right)\lambda_y\lambda_z}{4\left(\lambda_y\lambda_z + \frac{u_yu_z\lambda_x^2}{u_x^2} + \frac{u_z\lambda_x\lambda_y + u_y\lambda_x\lambda_z}{u_x}\right)} \tag{6.10}$$

According to the above proposed, the conclusion that $R(O)$ multiplied by the formula (6.1) can be obtained after the improvement of longitudinal collision risk probability value P_x':

$$\begin{aligned} P_x' &= 2 \times E(O) \times L(S) \times P_y(0) \times P_z(0) \times \left(1 + \frac{2u_y\lambda_x}{2u_x\lambda_y}\right) \\ &\quad \times \left(1 + \frac{2u_z\lambda_x}{2u_x\lambda_z}\right) \times R(O) \\ &= 2 \times E(O) \times L(S) \times P_y(0) \times P_z(0) \times \left(1 + \frac{2u_y\lambda_x}{2u_x\lambda_y}\right) \\ &\quad \times \left(1 + \frac{2u_z\lambda_x}{2u_x\lambda_z}\right) \\ &\quad \times \frac{\dfrac{3(u_z\lambda_x\lambda_y + u_y\lambda_x\lambda_z)}{u_x} + \left(\dfrac{1}{2}\pi + \dfrac{7}{4}\right)\lambda_y\lambda_z}{4\left(\lambda_y\lambda_z + \dfrac{u_yu_z\lambda_x^2}{u_x^2} + \dfrac{u_z\lambda_x\lambda_y + u_y\lambda_x\lambda_z}{u_x}\right)} \end{aligned} \tag{6.11}$$

According to the formula (6.1) (Wu), the impact of longitudinal collision risk value of the main factors through the longitudinal layer between the probability of aircraft, the following analysis of the calculating methods for them.

Compared to the radar monitoring technology, ADS-B technology has a great improvement in navigation and monitoring performance, so the calculation of the aircraft longitudinal spacer through rate is made mainly from the two aspects of the navigation and monitoring of airborne equipment. That is,

$$L(S) = L_n(S) \times L_s(S) \tag{6.12}$$

The position information of ADS-B system is determined from the satellite navigation system (GNSS), and most of the GNSS use the GPS system in the United States. ADS-B under the control of each aircraft according to the location information of airborne GPS through the transponder transmits its own position and other aircraft position information by airborne

ADS-B display to avoid conflicts between aircrafts. The position error of GPS follows the mixed Gauss distribution as follows:

$$f(x) = (1-\eta)N(x|a_1) + \eta N(x|a_2),\ \ 0<\eta<1 \tag{6.13}$$

where $(1-\eta)$ is the general error ratio, η is the impact of the collision risk of a large proportion of the error, a_1 is the general error probability density function corresponding to the parameters, and a_2 is the large error probability density function of the corresponding parameters.

When the aircraft flying in the RNP-n program meet the following conditions:

$$\int_{-n}^{n} \frac{1}{2a_1}\exp\left(-\left|\frac{x}{a_1}\right|\right)dx = 95\% \tag{6.14}$$

When n is determined, you can get them a_1.

Assuming that two aircraft positions are independent of each other, then the longitudinal cross ratio based on the navigation error is calculated by convolution integral:

$$L_n(S) = 2\lambda_x \int_{-\infty}^{+\infty} [f(x)f(S-x)]dx \tag{6.15}$$

Based on the error model of the monitoring, the model of the yaw error is subject to exponential distribution:

$$g(x) = \frac{1}{2b_1}\exp\left(-\left|\frac{x}{b_1}\right|\right) \tag{6.16}$$

Because the refresh rate and reaction time of the airborne receiver, the monitor, and the monitor system of the ground station are the main reasons for the yaw of the aircraft, assuming that the monitoring performance parameter is m, $m = (\text{Refresh time} + \text{Reaction time}) \times u_x$. The position error of the aircraft is therefore satisfied by the following expressions:

$$\int_{-m}^{m} \frac{1}{2b_1}\exp\left(-\left|\frac{x}{b}\right|\right)dx = 95\% \tag{6.17}$$

The longitudinal traverse rate based on the monitoring error is

$$L_s(S) = 2\lambda_x \int_{-\infty}^{+\infty} [g(x)g(S-x)]dx \tag{6.18}$$

So, the formula (6.15) and the formula (6.18) into the formula (6.12) can be obtained through the aircraft longitudinal spacer probability.

$$\begin{aligned} L(S) &= 4\lambda_x^2 \int_{-\infty}^{+\infty} [f(x)f(S-x)]dx \int_{-\infty}^{+\infty} [g(x)g(S-x)]dx \\ &= 4\lambda_x^2 F(S)G(S) \end{aligned} \tag{6.19}$$

where

$$\begin{aligned} F(S) &= \int_{-\infty}^{+\infty} [f(x)f(S-x)]dx \\ G(S) &= \int_{-\infty}^{+\infty} [g(x)g(S-x)]dx \end{aligned} \tag{6.20}$$

So, the formula (6.15) and the formula (6.18) into the formula (6.12) can be obtained through the aircraft longitudinal spacer probability.

6.1.2 Path Planning Based on PCNNs

In this section, a path model is designed for path planning. In addition, many variables are defined and used in this section.

Path architecture

The path architecture is proposed as follows

Each node in the space is connected only to its immediate neighbors, which constitute a subset $\Re_i$ in Ω, as shown in Fig. 6.6, generated by the pulse wave propagation at a constant speed. In this way, travel time between neurons is only proportional to the path length between them. Complexity of the algorithm is only related to the complexity of the shortest path length and map, and it does not require any prior knowledge about the target and obstacle motion and does not need to do a

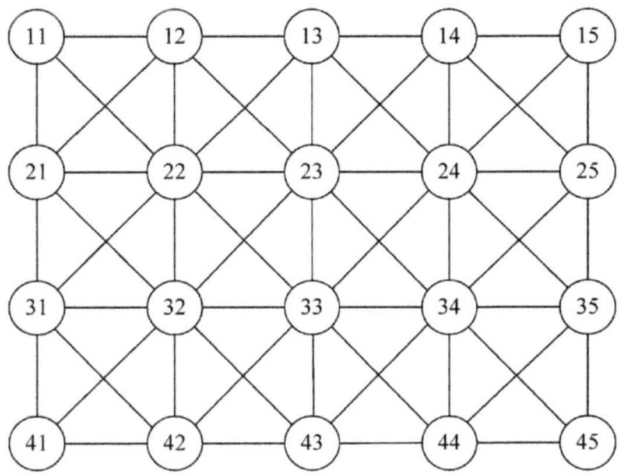

Figure 6.6 Networks structure.

cost–function optimization. In addition, there is not any comparison computation within the wave propagation of each neuron. Similar to others who have been proposed that the shortest path-planning algorithm needs the entire current information environment, including position of the robot, obstacles, and the target point.

6.1.3 Network Architecture

In the network architecture, all the nodes and edges are designed neurons, then the neuron j is designed as y_j and the connection weight from the neuron j to the neuron i is defined as x_{ij} in Fig. 6.7.

The PCNN is proposed as

$$\begin{cases} \dot{x}_{ij} = a_{ij}(x_{ij} + y_i)(1 - y_j) \\ y_j = \text{step}\left(\max\limits_{i \in R_i}\{x_{ij}\} - 1 \right) \end{cases} \tag{6.21}$$

where the x_{ij} is the path form neuron i to neuron j where,

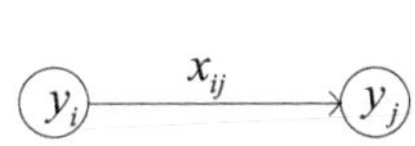

Figure 6.7 Neurons of nodes and edges.

$$a_{ij} = \frac{\ln 2}{s_{ij}} \tag{6.22}$$

For the neural networks state,

$$x_{ij}(0) = 0, y_j(0) = 0 \tag{6.23}$$

The proposed model contains nodes and directed edges, in which the nodes express the grid location and path between two nodes.

Before discussing the model itself, some notations and definitions about neuron firing and the firing time are defined as follows.

The node neuron j is represented by y_j such that

$$\begin{aligned} y_j(t) &= \text{step}\left(\max_{i \in R_i}\{x_{ij}(t)\} - 1\right) \\ y_j(0) &= 0 \end{aligned} \tag{6.24}$$

The edge from node i to node j is represented by x_{ij} such that

$$\begin{aligned} \dot{x}_{ij}(t) &= a_{ij}(x_{ij}(t) + y_i(t))(1 - y_j(t)) \\ x_{ij}(0) &= 0 \end{aligned} \tag{6.25}$$

where t is time, the neuron j is stimulated only from its neighbor node neuron i, $i \in \Re_i$ through edge neuron x_{ij}.

For any given node neuron j, the first node neuron in its neighbor set $\Re_j$ to fire is denoted as $\Re_j^F$ and the firing time of this neuron is denoted as $t_{\text{fire}}^{\Re_j^F}$.

After node neuron i fired, and $i \in t_{\text{fire}}^{\Re_j^F}$, then the edge neuron x_{ij} is stimulated, when the state of the edge neuron $x_{ij} = 1$, then the node neuron j is fired.

The firing time of the neuron j is denoted as t_{fire}^j.

A node neuron j will not fire only if some node neuron j in its neighbor set fires.

Node neuron j and $j \in \Re_i$ fires at a time t_{fire}^j; if node neurons k and $k \in \Re_i$ in the same neighborhood fires at time t_{fire}^k, then the edge neurons x_{ij}, x_{ki} are stimulated at t_{fire}^i t_{fire}^k, respectively, and neuron x_{ki} could reach fire before neuron x_{ji} fire, then node neuron i is fired; when $x_{ki}(t_i) = 1$ and $x_{ji}(t_i) < 1$, then the neuron x_{ji} is not active, so $x_{ji}(t) = x_{ji}(t_i) < 1, t > t_i$.

6.1.4 Path Planning Based on Neural Networks

Theorem 6.1: For any given node neuron i, j, k, edge neuron is x_{ij}, where $i, k \in \Re_j$, and distances between nodes i, j, s_{ij}. Suppose neuron i fires at time t^i_{fire} and that j will fire at some future time t^j_{fire} and satisfy the

$$t^j_{\text{fire}} = t^i_{\text{fire}} + s_{ij} \tag{6.26}$$

If the node neuron k is not fired before time t^j_{fire}, it would not fire in future.

Proof: If the neurons structure and algorithm are as (6.21), then the node neuron is i, j, k and $i, k \in \Re_j$.

The neuron i fires at time t^i_{fire}, based on the algorithm (6.24), then

$$y_i(t) = \begin{cases} 1, t > t^i_{\text{fire}} \\ 0, t < t^i_{\text{fire}} \end{cases} \tag{6.27}$$

The edge neuron x_{ij} satisfy the (6.25), then $x_{ij}(t) = 0$ when $t < t^i_{\text{fire}}$, then

$$\dot{x}_{ij}(t) = \frac{\ln 2}{s_{ij}}(x_{ij}(t) + 1) \tag{6.28}$$

When $t > t^i_{\text{fire}}$, then we have

$$x_{ij}(t) = e^{\frac{\ln 2}{s_{ij}}\left(t - t^i_{\text{fire}}\right)} - 1 \tag{6.29}$$

Then, $x_{ij}\left(t^j_{\text{fire}}\right) = 1$, $t^j_{\text{fire}} = t^i_{\text{fire}} + s_{ij}$ and when $t = t^j_{\text{fire}}$, $y_j(t) = 1$ and $\dot{x}_{ij}(t) = 0$, after $t > t^j_{\text{fire}}$, so $x_{ij}(t) = 1$, when $t > t^j_{\text{fire}}$.

This completes the proof of Theorem 6.1.

In fact, Theorem 6.1 shows that each neuron in the system fires just once upon the stimulation of a previously fired neighbor. This behavior is similar to the propagation of a wave. Smaller connection weights lead to earlier firing times. Thus, the proposed network can be used to solve some shortest path problems.

Assume that there is plane A with inherent path in air route and then the plane B can make the path planning to avoid the plane and get the target. Based on the collision model, the safety distance for plane B with plane A can be determined by Theorem 6.1, then the Theorem 6.2 shows the shortest path in safe fly.

Assume the plane A and plane B in the same air route, and real-time position $A(t)$, where $A(t) \in \Re$, then there exist dynamic circular region $D_A(t)$ with center at $A(t)$ and radius equal to safety distance. y_i^d indicates whether the neuron i in circular region $D_A(t)$, as $y_i^d(t) = 1$, if $i \in D_A(t)$, then $y_i^d(t) = 0$, if $i \notin D_A(t)$.

Theorem 6.2: Assume that there is plane A in the same area air route, then the plane B can make the path planning by following path-planning algorithm:

$$\begin{cases} \dot{x}_{ij} = a_{ij}(x_{ij} + y_i)(1 - y_j) \\ y_j = \text{step}\left(\max\limits_{i \in R_i} \left\{ \left(1 - y_j^d\right) x_{ij} \right\} \right) \end{cases} \tag{6.30}$$

where y_j^d is the safety node state.

Proof: If the neuron structure and algorithm are (6.30), after the dynamic neuron i fires at time t^i_{fire}, there exist two situations.

When $j \notin D_A(t)$, then $y_j^d(t) = 0$, the algorithm is same as in Theorem 6.1.

When $j \in D_A(t)$, then $y_j^d(t) = 1$ because $y_j = \text{step}\left(\max\limits_{i \in R_i} \left\{ \left(1 - y_j^d\right) x_{ij} \right\} \right)$; therefore, $y_j = 0$ indicates the neuron j would not fire.

This completes the proof of Theorem 6.2.

6.1.5 Simulation Example

To study neural network path-planning algorithm, the simulation is proposed to solve safety air route.

The aircraft A from node 11 to node 45.

From the simulation in Fig. 6.8–6.13, the neurons fired in proper order, and in Fig. 6.14, the plane A and plane B in their own air route maintain safety distance. In the simulation, the node 45 is fired at $t = 5{,}828$, from node 11 fired, and then, node 22 fired through edge neuron 1122, as 11-22-23-24-35-45 in Fig. 6.14.

6.2 ADS-B FOR MULTILATERATION SYSTEM USING BP NEURAL NETWORK

6.2.1 Multilateration

The multilateration (MLAT) system making full use of the existing equipment can realize a low cost as a new airport scene monitoring technology. The size of MLAT system receiving

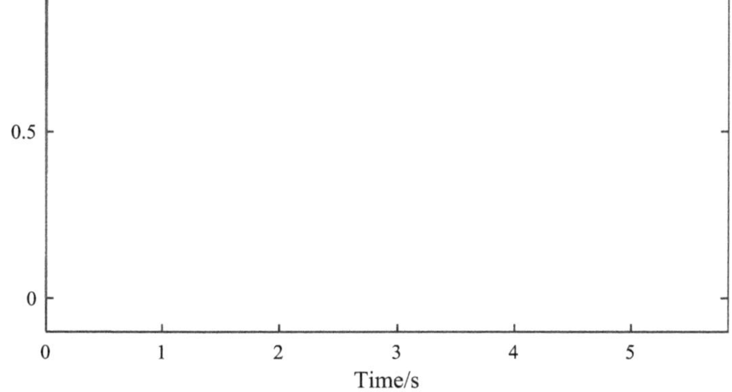

Figure 6.8 Neuron y_{11} fired at $t = 0$.

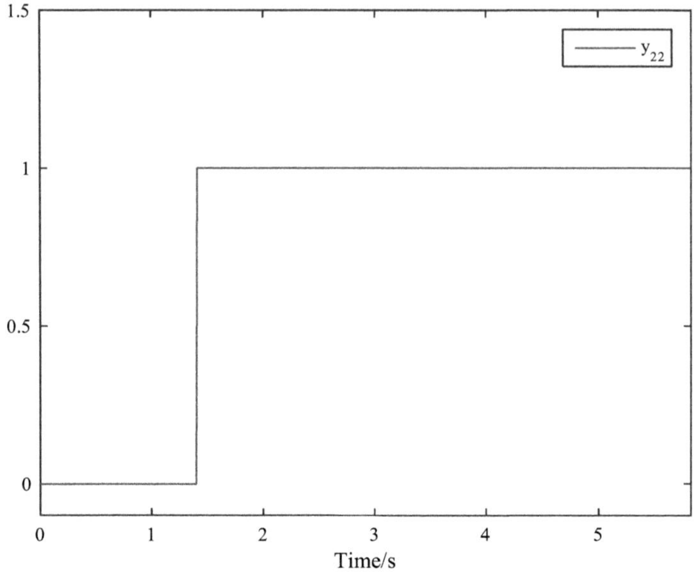

Figure 6.9 Neuron y_{22} fired at $t = 1.414$.

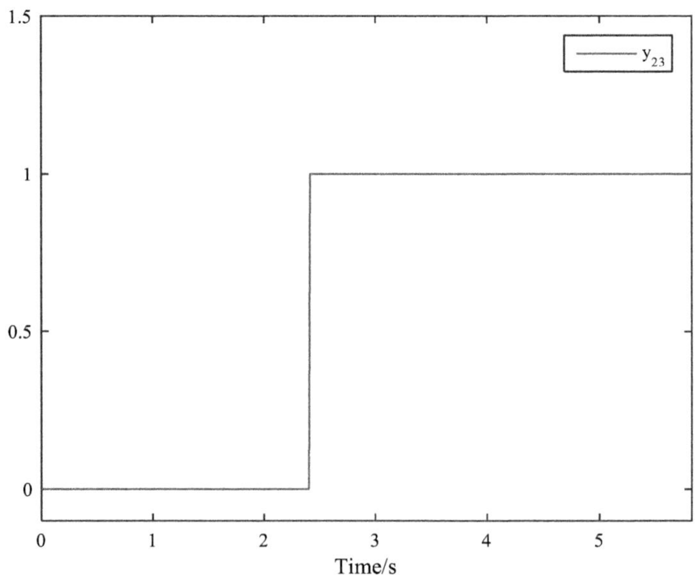

Figure 6.10 Neuron y_{23} fired at $t = 2.414$.

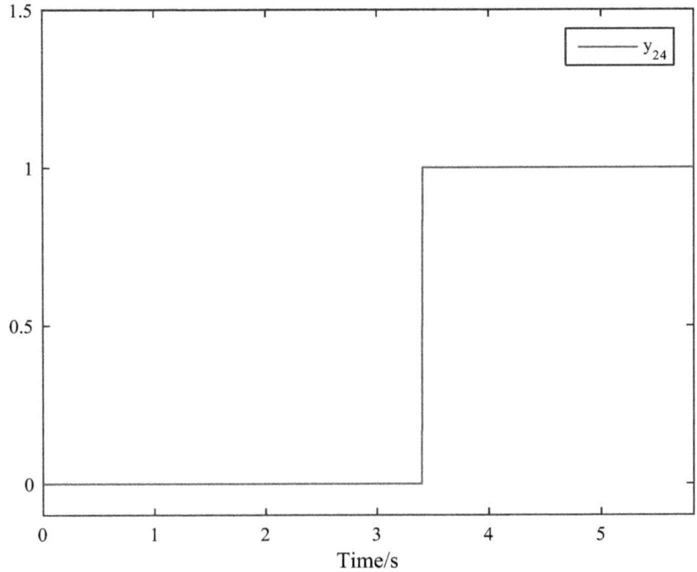

Figure 6.11 Neuron y_{24} fired at $t = 3.414$.

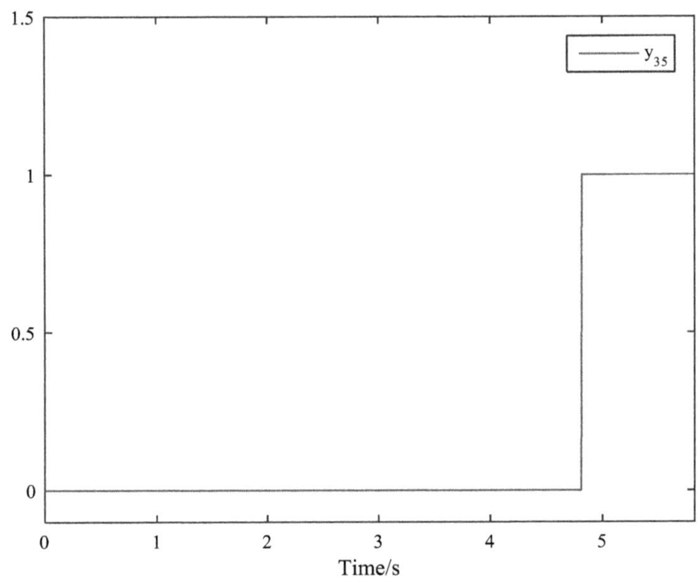

Figure 6.12 Neuron y_{35} fired at $t = 4.828$.

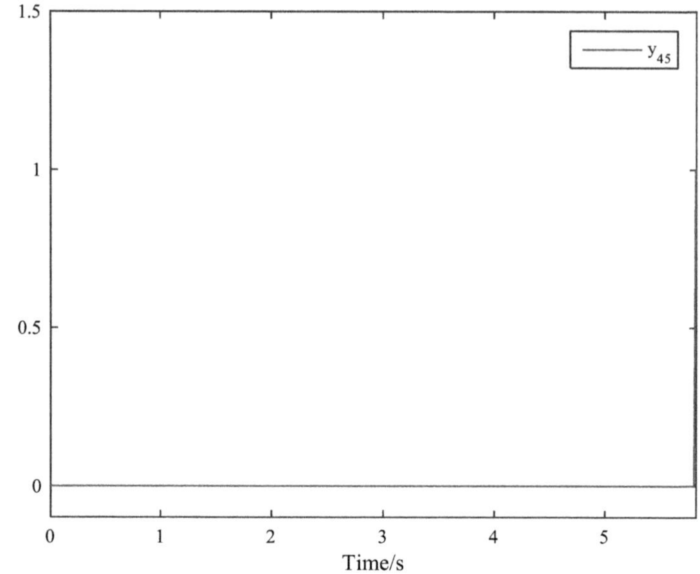

Figure 6.13 Neuron y_{45} fired at $t = 5.828$.

equipment is small and has flexible configuration. Multilateration has stronger system layout and coverage adaptability. In addition, the single base station promotes easily ADS-B base station, which reduces the cost of the ADS-B construction in the future. The MLAT system also provides reliable data for

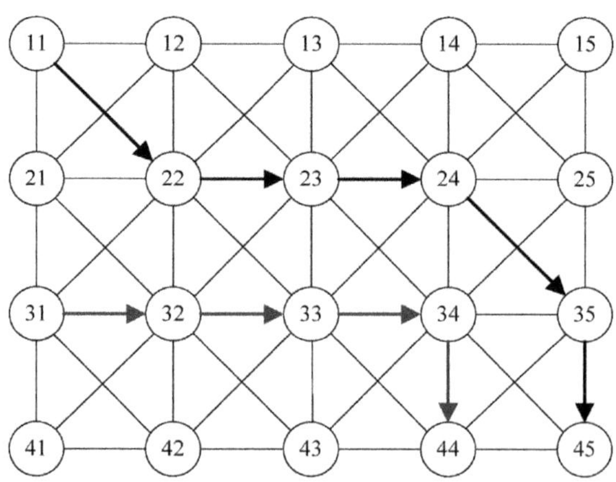

Figure 6.14 Trajectories of the plane *A* and plane *B*.

other target system, such as tower display system and airport base surveillance system. Multilateration could ensure the surveillance system to comprehend the accurate and real-time situation of the aircraft, especially in low visibility weather. So, the MLAT system greatly improves the ability to control and flight safety level (Lee et al., 2014; Tongjuan Zhao, 2018). Nowadays, the MLAT has been widely used in many parts of the world (Medina et al., 2012); it becomes core technology in the advanced scene activity guidance and control system (A-SMGCS) proposed by the International Civil Aviation Organization.

ADS-B (Reck et al., 2011) as a promising technology could improve air traffic surveillance level and extend existing RADAR-based surveillance mechanisms. The technical principle of ADS-B is shown in Fig. 6.15. In the radar coverage area, ADS-B base equipment network can be used as an auxiliary system for radar surveillance network, and it provides not less than radar interval standard ATC services. The ADS-B

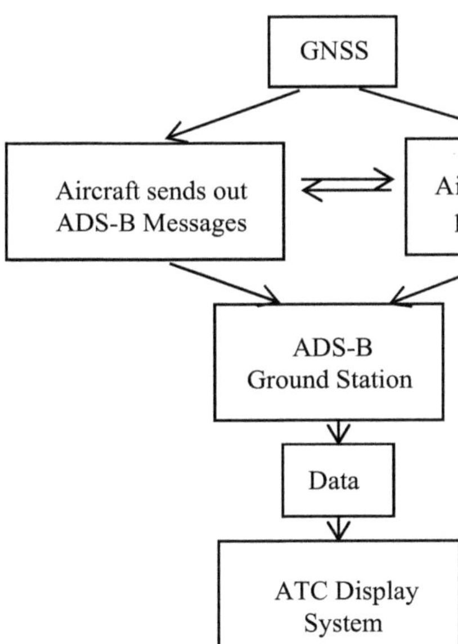

Figure 6.15 The principle of ADS-B technique.

transmitter is shown in Fig. 6.16. To improve the level of air traffic flow management, the use of uplink data broadcast ADS-B is necessary, which can provide all kinds of information services for the operation of the aircraft. ADS-B technology in the application of air traffic control indicates that the traditional air traffic monitoring technology is about to be a major change.

Multilateration is a technique that estimates the location of a wireless emitter, and the target position is determined by the time difference between signals arriving at multiple base stations in MLAT system (Linares et al., 2014). Therefore, MLAT system could reuse the existing infrastructure and does not require modification of ADS-B frame format. The construction of MLAT system is as Fig. 6.17. Thus, it realizes the injection of messages based on the ADS-B monitoring technology without any necessary upgrade to the aircraft transponder (Liu and Yang, 2011; Galati et al., 2010; Lewis et al., 2016). At present, the Capital International Airport in Beijing has successfully used the existing equipment to upgrade to MLAT system.

Figure 6.16 ADS-B transmitter.

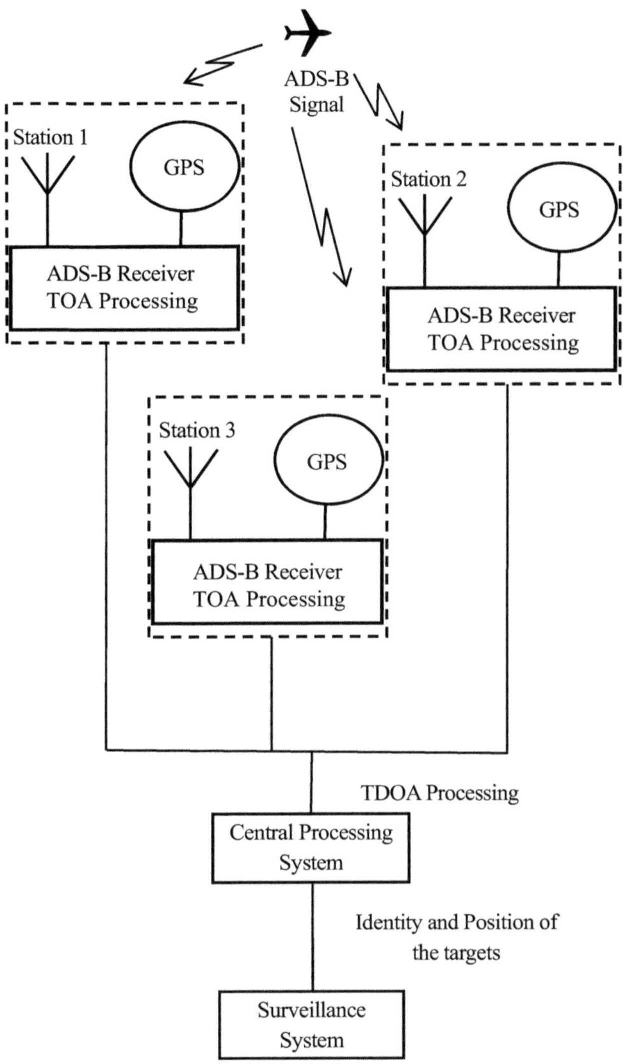

Figure 6.17 MLAT system—based ADS-B signal.

6.2.2 Multilateration System Theory and Positioning Algorithm

Multilateration, also called hyperbolic positioning, is the process of locating the object by the precise calculation of the arrival time difference of the signal (TDOA) (Romero and Mason,

2011) to different base stations. The theory of TDOA is shown in Fig. 6.18.

State of the MLAT system requires high-bandwidth communication lines for the correlation of signals from two base stations. Furthermore, the local clocks of the two stations must be synchronized with a high precision to make the arrival time difference more accurate.

Normally, MLAT system is composed of four or more base stations. The signals arriving from two base stations form a set of arrival time difference (Aguado et al., 2012). If the target is in the same level as the base stations, three base stations can calculate the target localization. But in most cases, the target and the base stations are not on the same level. Therefore, MLAT system requires at least four base stations to achieve accurate positioning of the target (Schwenke et al., 2008; Wang et al., 2011).

According to the theory as MLAT system, localization algorithm based on TDOA is widely used for target position (Kannan et al., 2007). The mathematical model of TDOA is as follows:

$$\begin{cases} R_0^2 = (x - x_0)^2 + (y - y_0)^2 + (z - z_0)^2 \\ R_i^2 = (x - x_i)^2 + (y - y_i)^2 + (z - z_i)^2 \\ R_{i,0} = R_i - R_0 = c\tau_{i,0} \\ i = 1, 2, 3 \end{cases} \tag{6.31}$$

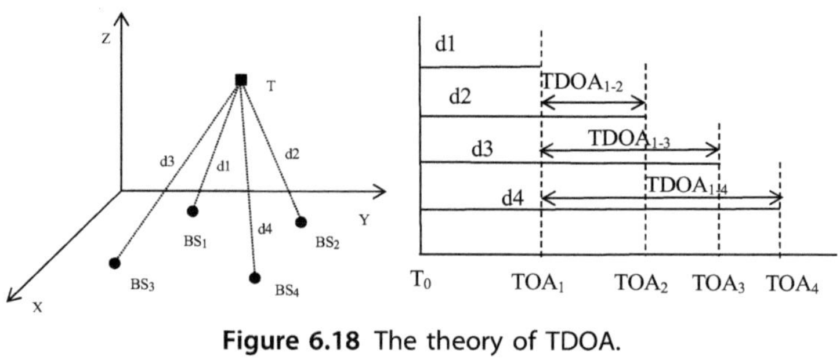

Figure 6.18 The theory of TDOA.

where (x, y, z) stands for the target location, (x_0, y_0, z_0) represents reference standard station to keep the clocks of other base stations in synchronization, and (x_i, y_i, z_i) represents the coordinates of the *ith* base station. R_0 and R_i, respectively, indicate the distance from the target to the reference standard station and the base station, $R_{i,0}$ stands for distance difference between the target to the reference standard station and *ith* base station. Where c represents speed of light and also represents the speed at which ADS-B signals propagate. The time difference between the reference standard station and the *ith* base station is $\tau_{i,0}$. Therefore, the TDOA principle, as shown in the above mathematical model, is the process of solving the above equation by locating the target of the MLAT system.

The algorithm with higher positioning accuracy can not only reduce the system computational error but also improve the efficiency of the MLAT system positioning. The Taylor algorithm does not require a higher channel environment but requires an initial prediction that is close to the target's real value. In the case of noise obeying the Gauss distribution, the Chan algorithm has higher positioning accuracy. When the channel environment is poor, Chan algorithm has a lower positioning accuracy, but it does not require an initial predictive value. Cater proposed that the target position could be searched from a large number of possible position sources. Although the method is more computational, its positioning accuracy is higher. To solve the problem of large computational complexity, Abel and Smith proposed a method of positioning, which can achieve higher accuracy in smaller error range, but this method is easy to fall into the local optimal solution. Therefore, the positioning algorithm with less calculation, high efficiency, and high precision is the core part of the MLAT system.

6.2.3 The Establishment and Process of BP Neural Network

Back propagation neural network algorithm was proposed by Rumelhart and McClelland et al., in 1986, which solved the learning problem of multilayer neural networks and greatly promotes the development of neural networks. BP neural network is the core part of the feedforward neural network and also the essence of the neural network system. It is widely used in the classification identification, approximation, regression, and other fields (Ondra et al., 2016). BP neural network is usually multilayer, which contain multiple hidden layers, input layer and output layer, and takes error back propagation learning methods. The hidden layer of BP neural network can be divided into one layer or multilayer.

BP neural network uses the error back propagation algorithm to study which divides the learning process to two stages. For the first stage, the data are propagated backward from the input layer to the hidden layer, and the network weights are trained in the direction of error reduction. The second stage is back propagation, if the output layer cannot get the expected value, the connection weights of the middle layer networks are trimmed before the output layer. With the continuous progress of learning, the final error is getting smaller until the output satisfies the expectation. The structure of BP neural network is shown in Fig. 6.19.

Suppose there has three layers of BP neural network containing a hidden layer, the number of input neurons, the hidden

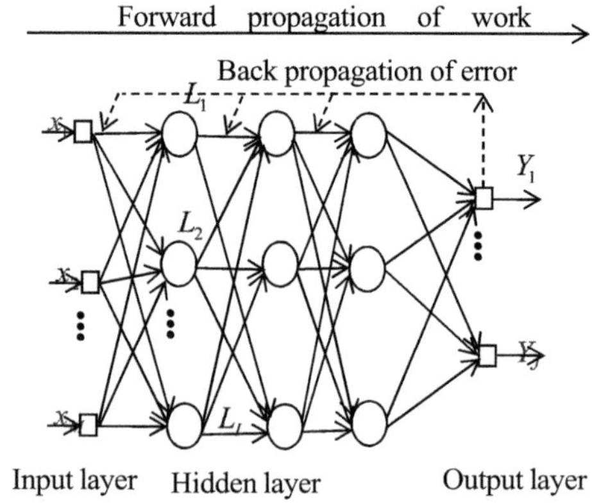

Figure 6.19 The structure of BP neural network.

layer, and output neurons are M, I, and J. The connection weight of the input layer to the hidden layer is ω_{mi}, between the hidden layer and the output layer ω_{ij}. u and v are the input and output of each layer.

According to the structure of three layer neural network, actual output $Y(n)$ and desired output $Y'(n)$, respectively, can be determined as

$$\begin{cases} Y(n) = \left[v_j^1, v_j^2, \dots v_j^J\right] \\ Y'(n) = \left[v_j^{1\prime}, v_j^{2\prime}, \dots v_j^{J\prime}\right] \end{cases} \tag{6.32}$$

where n represents the number of iterations, v_j^1 represents the output of the first neuron in the J layer, and the output of the input layer is equal to the input signal of the whole network:

$$v_M^m(n) = x_m(n) \tag{6.33}$$

Suppose error signal is $E_j(n) = Y'(n) - Y(n)$, define the error function $E(n)$ in the following equation:

$$E(n) = \frac{1}{2}\sum_{j=1}^{J} E_j^2(n) \tag{6.34}$$

1. Network initialization

   Select three layer network structures to train, according to the definition of the symbol as above to determine the BP network structure. Then, initialize threshold of the hidden layer and the output layer a and b.
2. Hidden layer output calculation

   According to the variables of input layer and the weights between input layer and hidden layer, calculate hidden layer output $v_i(n)$ as

$$v_i(n) = f\left(\sum_{m=1}^{M} \omega_{mi}(n)x_m(n) - a_i(n)\right) \quad i = 1, 2, \dots, I \tag{6.35}$$

where f represents the excitation function of the hidden layer, which has many different forms, here, the functions is selected as

$$f(x) = \frac{1}{1 + e^{-x}} \tag{6.36}$$

3. Output layer output calculation

According to the output of hidden layer v_i and connection weight ω_{ij} and threshold value b, calculate the actual output of BP neural network:

$$v_j(n) = \sum_{i=1}^{I} v_i(n)\omega_{ij}(n) - b_j(n) \quad j = 1, 2, \ldots, J \tag{6.37}$$

4. Error calculation

According to the actual output $Y(n)$ and desired output $Y'(n)$, calculate the prediction error of BP neural network $E_j(n)$:

$$E_j(n) = Y'(n) - Y(n) \quad j = 1, 2, \ldots, J \tag{6.38}$$

5. Weight updating

According to network prediction error, update weights ω_{mi} and ω_{ij}:

$$\omega_{mi}(n+1) = \omega_{mi}(n) + \eta v_i(n)(1 - v_i(n))x_m(n)\sum_{j=1}^{J} \omega_{ij}(n)E_j(n)$$
$$m = 1, 2, \ldots, M \;\; i = 1, 2, \ldots I \tag{6.39}$$

$$\begin{aligned} &\omega_{ij}(n+1) = \omega_{ij}(n) + \eta v_i(n)E_j(n) \\ &i = 1, 2, \ldots, I \;\; j = 1, 2, \ldots, J \end{aligned} \tag{6.40}$$

where η is learning rate.

6. Threshold updating

Update threshold a, b based on the network prediction error $E_j(n)$:

$$a_i(n+1) = a_i(n) + \eta v_i(n)(1 - v_i(n))\sum_{j=1}^{J} \omega_{ij}(n)E_j(n) \quad i = 1, 2, \ldots, I \tag{6.41}$$

$$b_j(n+1) = b_j(n) + E_j(n)$$
$$j = 1, 2, \ldots J \tag{6.42}$$

7. Threshold updating

Judge the satisfaction of constraint condition. If satisfied, output the updated weight and threshold. Instead, return steps (2).

(Wu) Forward propagation of the work signal

In the BP neural network, the work signal starts from the input layer and reaches the output layer through the hidden layer, and the total error of the system is finally obtained.

In the three layer of BP neural network, the output of the input layer is calculated first. The output of each input layer neuron is equal to the input signal of the network. The input of the neurons in the hidden layer is equal to the weighted sum of the neurons input from the previous layer. Therefore, the output of the hidden layer neurons is

$$v_I^i(n) = f\left(\sum_{m=1}^{M} \omega_{mi}(n) v_M^m(n)\right) \tag{6.43}$$

When a hidden layer is added, as shown in Fig. 6.19, the formula for calculating the output of the hidden layer is

$$\begin{cases} v_I^1(n) = f_1\left(\omega_{(x1)1}(n)x_1(n) + \omega_{(x2)1}(n)x_2(n)\right) \\ v_I^2(n) = f_2\left(\omega_{(x1)2}(n)x_1(n) + \omega_{(x2)2}(n)x_2(n)\right) \\ v_I^3(n) = f_3\left(\omega_{(x1)3}(n)x_1(n) + \omega_{(x2)3}(n)x_2(n)\right) \\ v_I^4(n) = f_4\left(\omega_{14}(n)v_I^1(n) + \omega_{24}(n)v_I^2(n) + \omega_{34}(n)v_I^3(n)\right) \\ v_I^5(n) = f_5\left(\omega_{15}(n)v_I^1(n) + \omega_{25}(n)v_I^2(n) + \omega_{35}(n)v_I^3(n)\right) \end{cases}$$

The input of the *j*th neuron of the output layer is equal to the weighted sum of the output of the previous hidden layer:

$$u_J^j(n) = \sum_{i=1}^{I} \omega_{ij}(n) v_I^i(n) \tag{6.44}$$

The output of the jth neuron of output layer is

$$v_J^j(n) = f\left(u_J^j(n)\right)$$

According to Fig. 6.20, the output of the output layer is

$$v_J(n) = f_6\left(\omega_{46}(n)v_I^4(n) + \omega_{56}(n)v_I^5(n)\right)$$

The error of the jth neuron in the output layer is

$$E_j(n) = v_J^{j'}(n) - v_J^j(n)$$

The total error of the system is

$$E(n) = \frac{1}{2}\sum_{j=1}^{J} E_j^2(n)$$

2. Back propagation of error signal
First, the weights ω_{ij} between the hidden layer and the output layer are adjusted. According to the gradient of the error to the ω_{ij}, adjust along the reverse:

$$\Delta\omega_{ij}(n) = -\eta\frac{\partial E(n)}{\partial \omega_{ij}(n)}$$

$$\omega_{ij}(n+1) = \omega_{ij}(n) + \Delta\omega_{ij}(n)$$

Decompose the gradient into a chain as

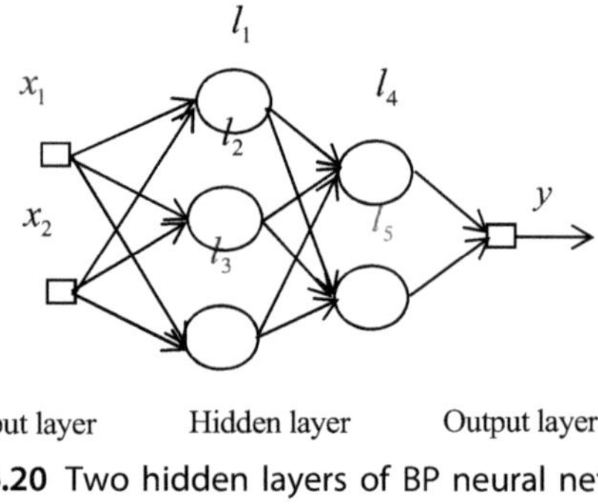

Figure 6.20 Two hidden layers of BP neural network.

$$\frac{\partial E(n)}{\partial \omega_{ij}(n)} = \frac{\partial E(n)}{\partial E_j(n)} \times \frac{\partial E_j(n)}{\partial v_J^j(n)} \times \frac{\partial v_J^j(n)}{\partial u_J^j(n)} \times \frac{\partial u_J^j(n)}{\partial \omega_{ij}(n)}$$

According to the previous definition, we can get

$$\begin{cases} \dfrac{\partial E(n)}{\partial E_j(n)} = E_j(n) \\ \dfrac{\partial E_j(n)}{\partial v_J^j(n)} = -1 \\ \dfrac{\partial v_J^j(n)}{\partial u_J^j(n)} = f'\left(u_J^j(n)\right) \\ \dfrac{\partial u_J^j(n)}{\partial \omega_{ij}(n)} = v_I^i(n) \end{cases}$$

So, the obtained gradient is

$$\frac{\partial E(n)}{\partial \omega_{ij}(n)} = -E_j(n) f'\left(u_J^j(n)\right) v_I^i(n)$$

The adjustment of weight is

$$\Delta\omega_{ij}(n) = \eta E_j(n) f'\left(u_J^j(n)\right) v_I^i(n)$$

In the output layer, the transfer function generally maintains a linear relationship, so its derivative is 1 as

$$f'\left(u_J^j(n)\right) = 1$$

So, it exists

$$\Delta\omega_{ij}(n) = \eta E_j(n) v_I^i(n)$$

Then, adjust the weights between the input layer and the hidden layer. According to the previous derivation, there are

$$\Delta\omega_{mi}(n) = nf'\left(u_I^i(n)\right) v_M^m(n) \sum_{j=1}^{J} E_j(n) f'\left(u_j^i(n)\right) \omega_{ij}$$

6.2.4 The Simulation of BP Neural Network

BP neural network, the basic structure of which is composed by nonlinear units, could approximate any function in theory. The target performing a flight mission is not in linear trajectory because of the uncertainty of flight trajectory and the probability of sudden events. Therefore, the nonlinear approximation of BP neural network can determine its trajectory very well. Therefore, it is of strong nonlinearity mapping.

The BP neural networks determine network structure and get desired output via regulating the learning method and the number of hidden layers and neurons in each layer. Besides, BP neural network has the characteristics of distributed processing of data to train a large number of experimental data simultaneously and speedup positioning. From the above, the BP neural network for position is proposed.

In the process of determining the structure of BP neural networks, using 500 to 1000 groups TDOA data to train the network. Set the network structure containing one input layer with three neurons, an output layer with one neuron, and three hidden layers with five neurons. The learning rate of the BP network is 0.01, and momentum parameter is 0.9. Set error vector as 0.01 m. Select target track range 0—3000 m. Select the main base station as (0, 0), other base station coordinates are (0, 30), (30, 0), and (30, 30). When the number of training reaches 2818 times, it meets the error requirements of the system setting, the simulation results are shown as Fig. 6.21.

Weights of the BP neural networks as follows:

The input layer to the first hidden layer

Weights	1.8984	−0.2349	−0.6535
	3.1281	−0.2599	0.8085
	−1.3781	0.5635	1.8820
	1.0401	1.3876	0.9777
	−0.7614	2.0633	2.9917

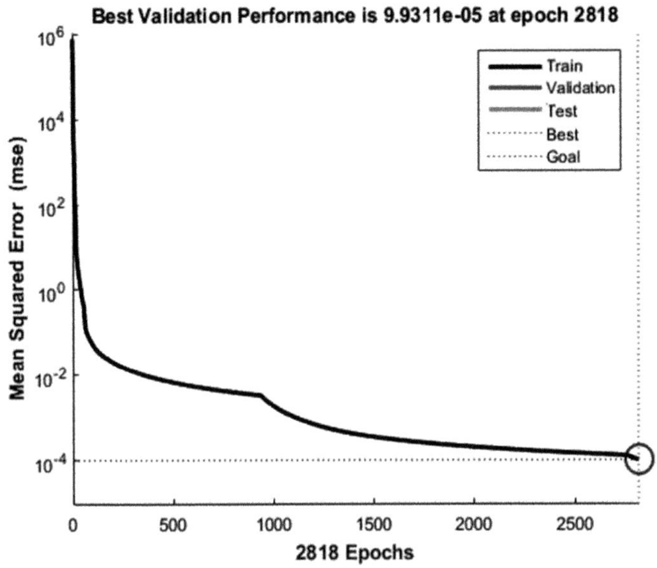

Figure 6.21 The simulation of BP neural network.

The first hidden layer to the second hidden layer

Weights	−1.0062	−0.2967	0.8538	0.3716	1.3729
	−0.1029	−1.1743	0.1463	−1.1445	0.1947
	1.1224	0.7290	−0.6171	−0.8566	−0.6599
	−1.3406	0.4183	−0.1916	−2.7689	1.6307
	0.8197	0.3323	0.1291	−0.5483	−1.4223

The second hidden layer to the third hidden layer

Weights	−0.4580	0.0032	−1.0535	0.7241	1.4065
	−0.2657	0.0647	−0.3769	2.5756	−0.9182
	1.7210	1.1840	0.4902	−0.0476	−0.1293
	0.7804	0.8750	−1.3930	0.7788	0.4075
	−0.6895	0.3040	−1.0299	−0.6993	1.2981

The third hidden layer to the output layer

Weights	1.3446	−1.6849	−0.4671	−1.2892	−0.0282

Thresholds of the BP neural networks as:

	The input to the first hidden layer	The first to the second hidden layer	The second to the third hidden layer	The third hidden to the output layer
Thresholds	−2.6991	1.8811	1.8978	1.1094
	0.0367	1.5798	0.7350	
	−0.3435	0.3555	0.2901	
	−0.0308	1.2168	0.7209	
	−0.7846	1.8884	−1.9856	

Finally, the output error is 9.9311e-05. The 2818th epochs error is shown as Fig. 6.21.

The simulation process is shown in Fig. 6.22.

Selected test data randomly in range of 0−3000 m, the actual output and the expected output as Figs. 6.23 and 6.24.

In this study, we described MLAT system with ADS-B signal and proposed the location algorithm based on BP neural network. A large number of experimental data are used to train

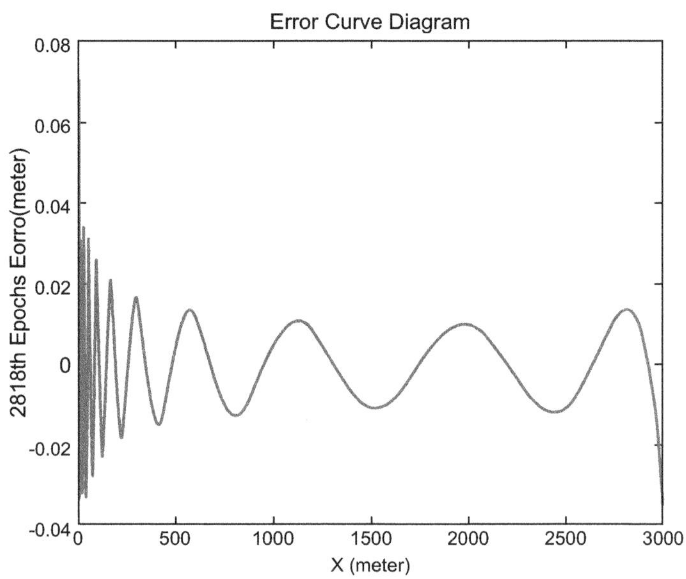

Figure 6.22 The 2818th epochs error of BP neural network.

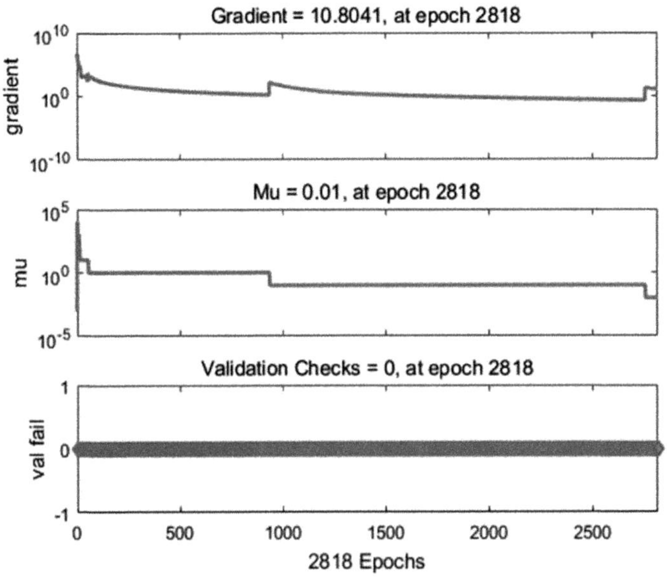

Figure 6.23 The simulation process for 2818 epochs of BP neural network.

the BP neural network to determine the network structure. The output error is set at 0.01 m to improve the positioning accuracy of the algorithm in theory. Finally, the actual output and

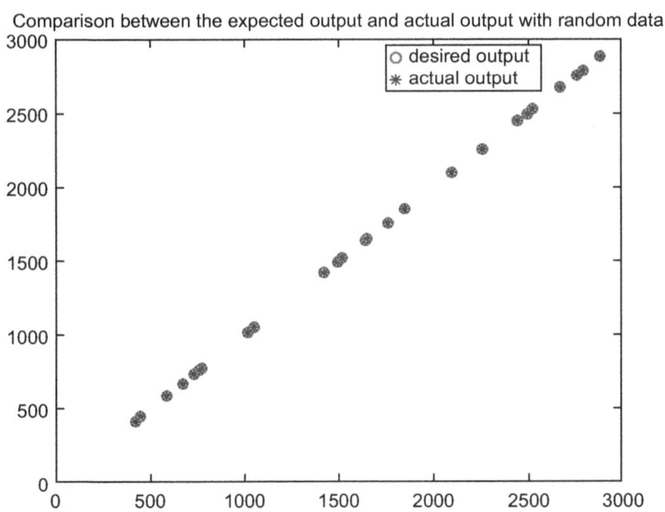

Figure 6.24 The simulation for test data randomly.

expected output are simulated by random selection of data, and the effectiveness of the algorithm is proved.

REFERENCES

Aguado, S., Samper, D., Santolaria, J., Aguilar, J.J., 2012. Identification strategy of error parameter in volumetric error compensation of machine tool based on laser tracker measurements. International Journal of Machine Tools and Manufacture 53, 160–169.

Ali, B.S., Ochieng, W.Y., Schuster, W., Majumdar, A., Chiew, T.K., 2015. A safety assessment framework for the Automatic Dependent Surveillance Broadcast (ADS-B) system. Safety Science 78, 91–100.

Besada-Portas, E., Torre, L.D.L., Moreno, A., 2013. On the performance comparison of multi-objective evolutionary UAV path planners. Information Sciences an International Journal 238, 111–125.

Cimino, A., Mirabelli, G., 2009. Modelling and simulation and ergonomic standards as support tools for a workstation design in manufacturing system. International Journal of Simulation and Process Modelling 5.

Espelos, N.J., Acosta, L., Alonso, D., 2013. Path planning approach based on flock dynamics of moving particles. Applied Soft Computing 13, 2159–2170.

Galati, G., Leonardi, M., Cavallin, A., Pavan, G., 2010. Airport surveillance processing chain for high resolution radar. IEEE Transactions on Aerospace and Electronic Systems 46, 1522–1533.

Gosciniak, I., 2015. A New Approach to Particle Swarm Optimization Algorithm. Pergamon Press, Inc.

Kannan, A.A., Fidan, B., Mao, G., Anderson, B.D.O., 2007. Analysis of flip ambiguities in distributed network localization. Information, Decision and Control 193–198.

Lee, S., Park, C., Min, J.L., Kim, S., 2014. Multihop range-free localization with approximate shortest path in anisotropic wireless sensor networks. EURASIP Journal on Wireless Communications and Networking 2014, 1–12.

Lewis, A.J., Campbell, M., Stavroulakis, P., 2016. Performance evaluation of a cheap, open source, digital environmental monitor based on the raspberry Pi. Measurement 87, 228–235.

Linares, J.M., Chaves-Jacob, J., Schwenke, H., Longstaff, A., Fletcher, S., Flore, J., Uhlmann, E., Wintering, J., 2014. Impact of measurement procedure when error mapping and compensating a small CNC machine using a multilateration laser interferometer. Precision Engineering 38, 578–588.

Liu, Y., Yang, Z., 2011. Location, Localization, and Localizability: Location-Awareness Technology for Wireless Networks.

Mccallie, D., Butts, J., Mills, R., 2011. Security analysis of the ADS-B implementation in the next generation air transportation system. International Journal of Critical Infrastructure Protection 4, 78–87.

Medina, D., Hoffmann, F., Rossetto, F., Rokitansky, C.H., 2012. A geographic routing strategy for North Atlantic in-flight internet access via airborne mesh networking. IEEE/ACM Transactions on Networking 20, 1231–1244.

Mohemmed, A.W., Sahoo, N.C., Tan, K.G., 2008. Solving shortest path problem using particle swarm optimization. Applied Soft Computing 8, 1643–1653.

Ondra, V., Sever, I.A., Schwingshackl, C.W., 2016. A method for detection and characterisation of structural non-linearities using the Hilbert transform and neural networks. Mechanical Systems and Signal Processing 83, 210–227.

Qiang Chen, Y.W., Zhongjun, H.U., 2018. Finite time synergetic control for quadrotor UAV with disturbance compensation. International Journal of Applied Mathematics in Control Engineering 1.

Reck, C., Reuther, M.S., Jasch, A., Schmidt, L.P., 2011. Independent Surveillance Broadcast — ADS-B Receivers with DOA Estimation.

Romero, L.A., Mason, J., 2011. Evaluation of direct and iterative methods for overdetermined systems of TOA geolocation equations. IEEE Transactions on Aerospace and Electronic Systems 47, 1213–1229.

Schwenke, H., Knapp, W., Haitjema, H., Weckenmann, A., Schmitt, R., Delbressine, F., 2008. Geometric error measurement and compensation of machines—an update. CIRP Annals - Manufacturing Technology 57, 660–675.

Sun, T.Y., Huo, C.L., Tsai, S.J., Yu, Y.H., Liu, C.C., 2011. Intelligent flight task algorithm for unmanned aerial vehicle. Expert Systems with Applications an International Journal 38, 10036–10048.

Tongjuan Zhao, J.W., 2018. Multimodal transport path planning based on multilateration (MLAT) system using a pulse neural network model. International Journal of Applied Mathematics in Control Engineering 1, 55–61.

Wang, D., Duan, Y., Wang, J., 2015. Environment exploration and map building of mobile robot in unknown environment. International Journal of Simulation and Process Modelling 10, 241–252.

Wang, X., Yang, Z., Luo, J., Shen, C., 2011. Beyond rigidity: obtain localisability with noisy ranging measurement. International Journal of Ad Hoc and Ubiquitous Computing 8, 114–124.

Wu, G., Qiu, D., Yu, Y., Pedrycz, W., Ma, M., Li, H., 2014. Superior solution guided particle swarm optimization combined with local search techniques. Expert Systems with Applications 41, 7536–7548.

Xie, C., Zhao, C., Dong, D., Zhong, P., 2014. Electric vehicle industry development environment evaluation in China based on BP neural network. International Journal of Simulation and Process Modelling 9, 234.

Zhang, Yanbo, 2011. Study of ADS-B data evaluation. Chinese Journal of Aeronautics 24, 461–466.

FURTHER READING

Wu, X.L., 2012. Synchronisation of unified chaotic systems with uncertain parameters in finite time. International Journal of Modelling, Identification and Control 17, 295–301.

INDEX

'*Note:* Page numbers followed by "f" indicate figures, "t" indicate tables.'

CPI Antony Rowe
Chippenham, UK
2018-12-03 18:29